Absolutely No Regrets

Roger Schofield

Crumps Barn Studio

Crumps Barn Studio LLP
Crumps Barn, Syde, Cheltenham GL53 9PN
www.crumpsbarnstudio.co.uk

Copyright © Roger Schofield 2015

First printed 2015

The right of Roger Schofield to be identified as the author of this work has been asserted by him in accordance with the Copyright, Designs and Patents Act 1988.

All rights reserved. No part of this publication whether text or photograph may be reproduced, stored in a retrieval system, or transmitted in any form or by any means, electronic, mechanical, photocopying, recording or otherwise, without the prior permission of the copyright owner.

ISBN 978-0-9931799-6-9

Printed in Great Britain.

For Joan, Karen, Michelle and all future generations of my family

CONTENTS

PROLOGUE	1
MY PARENTS	2
EARLY YEARS	4
LIVING IN ESSEX	14
APPRENTICESHIP	31
SERIOUS WORK BEGINS	52
MARRIED LIFE	55
MONSANTO	63
APRIL COTTAGE	75
GLOBAL ENGINEERING	85
GENESIS	93
OUR EMPTY NEST	100
MOVING TO THE COTSWOLDS	103

PROLOGUE

I started to investigate my family tree in 1991 whilst working close to Somerset House (London) where various family records were kept at that time. It was fairly easy to discover the names of my ancestors, where they had lived and, sometimes, their professions. However, they were mostly just names and, unless I was very lucky, I couldn't find out anything about their daily lives, which was a shame (for me, personally). I, therefore, decided to record some details of my life so that future generations can, if they want, see what my life was like in the late 20th and early 21st Centuries.

MY PARENTS

My parents met during the 2nd World War whilst my father, Alan, was stationed in East Acton, London. He met my mother, Doreen Sadler, at a social club where, upon being seen looking at a sweet young lady, it is alleged he said, "It's your eyes, ducky, not the powder on your nose." Quite a novel chat up line but it worked, and my mother did have lovely deep brown eyes! They were married on 21st February 1942 at St Dunstan's Church, East Acton. My grandmother, Bertha, had wanted her son to marry the girlfriend he had had in Oldham prior to the war, so the relationship between her and my mother

Alan Schofield and Doreen Sadler

MY PARENTS

didn't get off to a good start!

My parents lived at Anton Villa, Simister Lane, Prestwich, Manchester after my father was demobbed from the army in 1945, and had returned to working as a clerk for the Co-op, with whom he had been employed before the war.

EARLY YEARS

I was born on 7th May 1947 in Boundary Park hospital in Oldham, Lancashire and returned to my parent's home in Prestwich soon after.

My father was transferred to Power Samus (part of the Co-op) in Holborn, London as a data processing clerk (a sort of early computer operator, using punched cards to enter the data) and so we moved to 28 Girton Close, Greenford, Middlesex in 1948 and my sister, Brenda, was born on 15th December 1949. The only memory I have from that period (and that's probably only from being reminded of it by my mother) is coming home from a pre-school nursery and collecting conkers. There were plenty to choose from because I was passing the horse chestnut tree before the school age children left school in the afternoon. My father was transferred again to an office in Watford in 1951 and so we moved to 33 Hazel Tree, Watford, where I started school. Sadly, I have no memories of living in Watford.

My father kept in touch with some of his army friends, one of whom owned a newsagent/tobacconist shop. He was about to buy a second shop and asked my parents if they'd run if for him. There was an unwritten agreement that my parents would only take out of the business what

EARLY YEARS

they needed for everyday living and that, eventually, they'd become part owners of the business. Our family therefore moved back to Manchester in 1953 to North Road, Longsight. My earliest memory from that time was that the shop next door sold and chips, and that it had a very high counter that I couldn't see over. When I visited, many years later, the counter was no higher than any other fish and chip shop but, of course, I wasn't very tall when I was six!

My parents used to get up very early every morning to get the newspapers ready for delivery. I desperately wanted

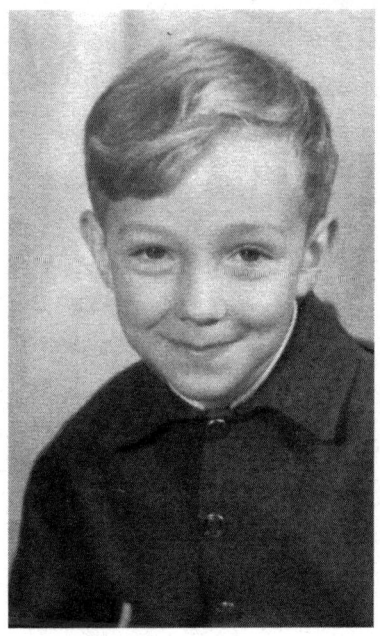

An early photograph

to be a paper delivery boy but it wasn't allowed because I wasn't old enough. I was promised that I could help another paper boy when I was older and perhaps do a paper round myself when I was 11. That was no good because I would have to wait for years! I did help out in the shop by getting things the customers wanted and then trying to add up the total cost in my head before Dad could ring up the amount in the cash drawer. I am sure that mental arithmetic helped my mathematics enormously.

We occasionally visited my father's parents, William and Bertha, at 7 Brierley Street, a small 'two up, two down' terraced house on a hill in the centre of Oldham. The house was owned by my grandparents and was typical of those in the area. It was entered via the front door directly from the street (there wasn't a front or rear garden but there was paved area at the back of the house) and had a living room and dining/kitchen downstairs and two bedrooms upstairs. It was a rare treat to go into the living room because it was kept tidy for special occasions. (What a pity when they were so short of space, but that was how things were done then). There was a toilet in the back yard but no flush mechanism! The toilet waste went straight into the sewer which was about ten metres below ground level. You can imagine what the smell was like! Toilet paper was a luxury that my grandparents couldn't afford so cut up squares of newspaper was the norm.

There wasn't a bathroom, so everyone washed at the kitchen sink, and used a tin bath in the kitchen once a week. (Privacy? What's privacy?).

My grandmother was a bit of a dragon and seemed to always have a scowl on her face and didn't seem to enjoy her grandchildren visiting! My grandfather had worked in the cotton industry in Lancashire and was now retired. I don't remember too much about him except he was short and thin, deaf and very short-sighted. He didn't have a lot to say but amused us by making animal noises. He also kept his slippers in a box by the fire so that they were kept warm.

I went to the local school about a half mile from the shop and remember the nicest girl in the class was called Jacqueline Taylor and she liked a boy called Robert Wild. (Why should I still be able to remember their names when I can't now remember the names of people I'm introduced to for more than a few hours? If only my memory could be made more efficient...). At playtime, all the boys would play football, Manchester United vs Manchester City. The game was very one-sided because you played for your favourite team. This meant that sometimes there were 30 boys playing for United and only ten playing for City. I usually played for City because I could get more involved in the game when playing in the smaller team. I became a United supporter after the Munich air crash in 1957 because I didn't like the fact that City played a league match on the Saturday after the crash, when most clubs postponed their games as a mark of respect. However, I didn't go to see them playing at Old Trafford whilst I was a boy, and it wasn't until 1963 that I went to West Ham with my father to see my team playing.

My first entrepreneurial venture was made at about the age of 7! I made a 'salt cellar' (a very simple origami shape) horoscope out of folded paper and wrote silly things inside like 'You will have sandwiches for tea' or 'You will fall over today'. Lots of kids were interested in it and so I started making and selling them for 1d (old money) each.

Guy Fawkes nights in Manchester were terrific fun because bonfires were built and lit in the middle of most side streets. The streets were concrete then so there was no damage to the road surfaces. The local treats around the bonfire were toffee apples, parkin (a ginger cake) and baked potatoes that were cooked in the bonfire embers. It was great to be able to gather around the warming bonfire whilst watching everyone's fireworks being lit and exploding into a riot of coloured lights. I remember returning home after a firework party to see Dad making something out of wood. Intrigued, I asked what he was making and he told me it was a fort that he was going to sell in the shop. I watched intently over many evenings to see how it was progressing and ultimately wishing it could be mine. Imagine my delight on Christmas day to find that same fort at the bottom of my bed! I had many happy times playing with that fort, its drawbridge and my toy soldiers.

Our family pet was a cat called Sooty which was black all over with white feet and we all enjoyed frequent cuddles and purring in our ears. He was great at catching mice that lived in the cellar and he wasn't supposed to go outside of the shop. Unfortunately, he managed to escape one day and didn't come back that night. Sooty returned the next morning but he had been badly hurt and, although the vet tried to help, he had to be put down. We didn't know how he got hurt but, from his injuries, I later concluded that he'd been severely kicked. We all missed him very much!

Whilst I don't remember being poorly, I guess I must have been fairly often because I was sent to a convalescent home in Conway, North Wales when I was seven. I had something called 'congestion of the lungs' which was apparently caused by infections due to a damp environment. I stayed in the home for about six weeks and had a great time! All of the younger boys, me included, lived in a dormitory and we had regular pillow fights. Most days we went to play on the sandy beach and dunes next to Conway Castle. I had regular visits from my family on Sunday afternoons when the shop was shut. I don't know why, but Brenda was never allowed to visit me and so she had to stay in the car whilst Mum and Dad took turns to be with me. When I returned home I remember thinking that our living room seemed very small in comparison to the large rooms at the home.

I don't remember being aware that Mum was pregnant but I do remember that on 14 July 1955 there was a fair in the park next to our school. Brenda and I stopped off at the fair on our way home from school and Nana Sadler (maternal grandmother) came to find us and take us home to see our new sister, Elaine. I don't remember how I felt about having another sister but I'm sure a brother would have been preferable to me at the time. The reality was that the 8 year difference in age meant that we had little in common during our childhood and I only got to really know Elaine when we were both adults.

We eventually moved from Longsight to Burnage, also in Manchester, because the owner of the shop wanted my parents to manage their other newsagent and tobacconist shop. Unfortunately, this second shop was closer to where the owner lived and it was easier for them to drop in to see how things were going. My parents saw this as unwarranted interference with the running of the shop.

It was whilst living in Burnage that we had the first holiday I can remember. We went to Scarborough for a week where I think we stayed in a caravan. My only memory of the holiday was going to a show where I, along with a number of other children, was invited up on to the stage. Whilst on the stage I needed to go to the loo but I didn't dare walk off so I just had to cross my legs.

EARLY YEARS

Unfortunately, we stayed on the stage too long and I had a little accident. How embarrassing!

When we returned to Manchester we found that the owner, who was looking after the shop whilst we were away, had completely reorganised the layout. My parents were furious and my father decided to look for alternative employment and leave the shop and its interfering owners. He was offered a job in London working for Allen & Hanburys as a manager of a data processing department and so it was time to leave Manchester. Needless to say, the share in the shop never materialised!

We had a "Standard 10" motor car at that time and we set off on the journey to London where we would be temporarily staying with my grandparents in Acton. We hadn't gone far when our dog, Rex, was sick in the car! We carried on but poor old Rex just didn't like travelling so we stopped near some boys and gave them money to take him to a dog's home. I don't suppose he got there but he may have been taken in by the boys' family.

It was August 1956 when we moved into 166 Western Avenue (the A40 main road into London from the West). My maternal grandfather, Frank Sadler, worked for Bronnley, making talcum powder and soap products. He had been a blacksmith for most of his working life but became disabled with Padgets disease and thus had to find a less physically demanding job. My grandmother, Hilda, was a housewife. I got on well with my grandparents, especially Nana, because I was her favourite (her first grandchild). My grandfather (we called him Pop-pop) had been a bit of a rogue in his early years and gambled too much for Nana's liking. One tale that was told about him was that he played cards on Christmas Eve and lost the money that was meant to be used to buy the Christmas vegetables so he raided an allotment on his way home to get the Brussels sprouts! He kept a bamboo cane next to the fire place and if he thought we had been naughty we would feel it across our backsides. In later years, Brenda decided to get rid of the cane and unsuccessfully tried to burn it. (I expect she felt the cane for that, too!).

My grandparents' house was rented from a company called Carter Pattersons (a road haulage firm) who had their storage depot at the end of the garden. However, it was quite large, compared with my other grandparents' house, and could easily accommodate us all for the six months we were there.

My Aunt Muriel, Uncle Arthur, cousins Doug, Carol and Sandra Kimber also lived in Acton and we saw quite a lot of them whilst we were in London. I went to the local school in September and on my first day was asked by a teacher what religion I was. I had no idea what religion was, let alone what branch I might belong to, so I asked the boy who sat next to me what he was and he said Catholic. I said "So am I" and he told the teacher. I knew I had got it wrong when the Catholic minority were segregated and sent to another classroom when the rest of the school went into the hall for assembly. I went with the majority for the morning service and later confirmed that I was not Catholic and had made a mistake.

LIVING IN ESSEX

About six months later my parents found a house in Chadwell Heath in Essex and we moved to 52 Burlington Gardens. Our new home was a three bedroom end-of-terrace house which had the main Southend to Liverpool Street railway line at the end of the garden. Steam trains were still using the line in 1958 and sparks from the engine would sometimes set fire to dry grass along the side of the railway line. Some trains would also make our house shake when they went past but, after we got used to it, they didn't keep us awake at night. Mum and Dad had a double bedroom and I had a small single room at the front of the house, whilst Brenda and Elaine shared a double room (and double bed) at the back, next to the bathroom. The water was heated by a 'back boiler' behind the fireplace in the 'living room' but that, obviously, only heated the water when a fire was alight. When there was no fire burning, water was heated in the kitchen and carried upstairs in saucepans and tipped into the wash basin or bath. Needless to say, we had baths in very shallow water. The only heating in the house was from this fireplace (central heating wasn't the norm in most houses) although we used a paraffin heater in the hall during the really cold winter months. It wasn't uncommon to have ice on the inside of our bedroom windows in the mornings and I used to get dressed in bed so I didn't get too cold! We also had a 'best' room at the front of the house that was rarely used but housed an upright piano that Dad would play at weekends. I had piano lessons for a while but didn't enjoy playing and gave up after a couple of years.

Just after we moved into our new home I went out into the street and saw some local boys playing with toy cars. I asked if I could play with them and was told an emphatic "No". The local boys were not very friendly and it was quite a shock to me. In Manchester, there would never be a problem joining a group of other children in their games even if you didn't know them. I eventually became accepted by a group that lived a bit further down our road and I ignored the unfriendly ones. I'd heard that people who lived in the South of England weren't very friendly and I'd experienced that soon after moving to Essex.

One of my favourite games was marbles. Any number of people could play and the object of the game was to push marbles with your finger to try and get them into a depression in a manhole cover. Any that went into the hole were won. I was very good at this game and won lots of marbles. I used to trade them back to the other boys in return for a ride on their push bikes. At that time I only had a tricycle and was desperately trying to learn how to ride a conventional bike. Once I could ride a two-wheeler, there was a chance I would get one for Christmas or my birthday.

Our car was sold soon after we moved to Chadwell Heath because the local railway station was only ½ mile from home and my father commuted to work by train to Bethnal Green, East London. Some evenings I'd like to walk on my own to the station to meet Dad when he got off the train. He used to walk very fast and I had to run just to keep up with him!

My local junior school was The Warren in Whalebone Lane, which was about ¾ mile from where we lived and I walked to and from school every day. My first class was with Mr Woodruff and I only remember that because our class football team was called Woodruff's Wanderers. I took my 11+ exam the following year but my result wasn't good enough to enable me to go to a grammar school so I moved up to the secondary modern part of The Warren. All new pupils of the senior part of the school took an entrance exam and I was put into the top class. I didn't get very good marks in the exams at the end of the first year so at the start of the second I decided to sit next to the boy who had come top. I suppose I thought I might be able to crib some of his work if I was sitting next to him. Anyway, the ploy worked and my marks improved.

My first job was as a newspaper delivery boy when I was 12 and some 6 six years after I'd first wanted to deliver papers in Manchester. I used to get up at about 5.30am to do my round before school, which wasn't too bad in the summer but, when it was getting colder and darker as winter approached, I found it difficult to get up. One particular morning when my alarm clock woke me up, I knew it was cold and could hear the rain pouring down, so I turned over and went back to sleep. When I went to the shop the following morning I found out that I had been sacked! It was a blessing in disguise as paper boys were not very well paid but I did need the money. My pocket money was only two shillings per week (10p in today's money). Another of my money-making ideas was to breed Chinchilla rabbits that had beautiful fur that could be used in the manufacture of fur coats. I bought a rabbit hutch and a rabbit I named Ena and was enthusiastic about my latest venture. As soon as Ena was old enough, she was introduced to a buck rabbit owned by the man I'd bought Ena from and a few weeks later I was the proud owner of six baby rabbits. These youngsters grew rapidly and, in no time at all, the hutch was too small and I was going to need more hutches. Mum and Dad weren't too keen on their garden being overrun with wood and mesh structures, and I was now having second thoughts about killing these sweet and cuddly creatures so plan 'B' was called for. I put a postcard in a local shop advertising pet rabbits for sale for five shillings each and quickly made a tidy profit! Ena had another litter later in the year but selling her offspring took longer this time and I soon lost interest in the venture. I did, however, keep Ena as a pet that Brenda and Elaine loved to hold and stroke.

I was desperately shy around girls and didn't know what to say to them even though I had two sisters. However, there was one girl that I liked very much and her name was Isabel Bisland. I decided to find out where she lived and so I waited outside of school and then followed her whilst walking on the other side of the road and at a discreet distance. The first part was in the direction of my home, although not the way I usually walked, but when she turned off I got cold feet and went directly home thinking that I'd try again another day. I saw her in the school playground the next day and, to my horror, she came up to me and said "Were you following me after school yesterday?" I quickly denied it and just said that I occasionally went a different way home. I had no inkling that I'd been spotted so my sleuthing wasn't as discreet as I'd thought. That put an end to me finding out where Isabel lived as it would have been too embarrassing to be caught out again!

In the second year at The Warren we took the 13+ exams, and got good enough marks to be invited to Romford Technical School for interview where they wanted to know what my work ambitions were. As my best school subject was 'technical drawing', I said I wanted to become a draftsman. I was eventually offered a place and moved there for the start of the 3rd year along with one other girl, Julia Tomlinson (we were the only ones to pass the 13+ from our class that year).

Romford Tech was about three miles from where we lived and I cycled every day, rain or shine, cold or hot, fog, rain or snow. All of the children who passed the 13+ were put into one class and were taught separately from the other kids of our age until we reached the 6th form. This was because we came from a number of different 'secondary modern' schools and we had been taught to different standards and needed to be brought to the same level as those who had passed the 11+ exams.

I was allowed my first pair of long trousers when I went to Romford but, amazingly, I was not the last boy to get some. There was one other boy, Jack Pitcher, who started at Romford still with short trousers! I was one of the smaller children in the class and was occasionally 'picked on' by bigger boys. It didn't really amount to bullying and I don't think it affected me in any adverse way. I'd sometimes retaliate but that usually meant I got a few extra bruises on my arm so my normal reaction was to walk away from trouble. On one occasion, I was having my lunch with classmates and the custard jug was empty by the time it was passed to me. As I went to the kitchen to get it refilled, I noticed Paul Haddon shaking the salt cellar over my dessert. There were spare salt cellars in the kitchen and I emptied the contents of one of them into my hand and, on my return to our table, dumped it into his stewed apple. He was not amused but the rest of the boys on our table thought it was hilarious.

Our Headmaster, Mr Mitchell (known as Mitch), was short, bald on top and had protruding teeth that affected his speech. A word he used often at our morning assemblies was 'others' when perhaps referring to boys fighting in the playground whilst 'others' were looking on. Unfortunately, because of his speech impediment, the word that came out of his mouth was 'udders' and this made lots of pupils collapse in laughter and giggles. Even some of the teachers who were sitting behind Mitch found it difficult to keep a straight face. Some of the more obvious gigglers were told to go to the headmaster's office to explain themselves before receiving several strokes of his cane!

I had not been taught French at The Warren and neither had about half of our class. Needless to say, it was very difficult for our French master, Mr Werner, to teach a class where half had two years knowledge of French and half had none. At the end of the first year I came close to bottom of the class in French and was extremely pleased that French was not compulsory in the next year.

My progress during that first year was not very good and I finished close to bottom of the class of 31. However, I dropped my worst subjects, French and History in the following year and, at the end of the 4th year, came 7th overall. There were quite a few pupils who opted for Applied Maths rather than French purely because it was 'the lesser of the evils'. Our maths master realised this and, rather than having the class disrupted by those who has no interest in the subject, split the class into two halves, those who wanted to learn the subject and those who didn't. The deal was that those who didn't could sit at the back of the class and do homework for other subjects as long as they were quiet. Brilliant idea! Mick Cork was someone who wanted to learn Applied Maths but wasn't very good at it and regularly achieved low marks for his homework. However, he always came to school with jammy sandwiches that he ate during mid-morning breaks. I didn't take any sandwiches to school and we struck a deal that he'd share his sandwiches with me in exchange for being able to copy my homework. At the end of term, our maths master was very impressed with Mick's progress when he achieved top homework marks beating me because I'd had a week away on holiday and thus missed a week's marks. The maths master was rather less impressed when Mick came bottom in the end of term exams! I took my 'O' level GCEs in 1963 and passed in six subjects, Maths, Applied Maths, English, Physics, Technical Drawing and Chemistry. I failed Geography but passed it during my next term. I achieved 'A' grades in Maths and Applied Maths and was awarded the school's

Mathematics prize.

Although I'd been listening to 'pop' music on the radio for some time, it wasn't until The Beatles released Please Please Me that my interest in their new style of music took off and I started buying my own 45rpm vinyl records. Thereafter, there were many groups being formed that played music that appealed to my generation, The Rolling Stones, The Who, The Hollies and The Yardbirds to name just a few. The BBC's 'Light Programme' didn't play much of the 'pop' music and so we needed to listen to Radio Luxembourg in the evenings to hear more; even though the quality of the radio reception was generally poor. Then, in 1964, pirate radio stations sprang up, with Radio Caroline being one of the most popular. They broadcast from ships moored in international waters just off the UK coast to circumvent the record companies' control of popular music and the BBC's radio broadcast monopoly. This 'new' music style was followed by a youth subculture known as Mods, who rode Lambretta and Vespa scooters and wore suits and clean cut outfits. The previous Rock and Roll music was followed by another subculture that became known as Rockers, who rode motorbikes and wore black leather jackets. These subcultures didn't get along and, especially on bank holiday weekends, they'd congregate at seaside towns for fights and rioting. I stayed well away from any trouble and merely enjoyed listening to the music with friends.

I always enjoyed sport at school but wasn't very good at football or cricket (although I did play once for the cricket second eleven). I was quite good at running but didn't particularly enjoy cross-country running, which was mandatory during the winter PE lessons when the football pitch was too muddy to play on. Once every year there was the house cross-country race where every boy was expected to run and gain 'house' points. I finished 6th overall, which was an excellent result for me and my house. Unfortunately, I then found myself selected for the school cross-country running team with matches against other schools on a Saturday afternoon. Our Geography teacher (Jim) owned a 1930s Rolls Royce and he used to take a few of us in it to these matches which was great fun. It was obviously a great honour for me to be selected to run for the school but it interfered with my Saturday job at the Fine Fare supermarket in Chadwell Heath. If I didn't work, I didn't get paid! I solved that problem by finishing a few races with poor results and got dropped from the team.

I'd been working at the local Fine Fare supermarket since I was 14 and worked for two hours on a Friday night and all day on Saturday for £1/week. I worked there for a couple of years and even got my friend Steven Holmes (he lived just along the road from us) a job also at weekends. The staff members at the supermarket were all friendly types and I soon became aware of the fiddles that went on.

The employees were allowed to buy damaged goods cheaply as they couldn't be sold to customers. This led to some staff deliberately denting tins or tearing packages of things they wanted so they could buy them cheaply at the end of the day. I enjoyed working in the supermarket under the manager, Mr Hasted, but the pay wasn't as good as in the Fine Fare in Romford where they earned £1.5 shillings for the same hours of work. Some of us 'Saturday boys' would ask the manager for a raise but he would just ask us if we were satisfied with our jobs. If we said 'No', he would say 'Then you had better look for another job' but if we said 'Yes, but the pay isn't very good' he would retort 'If you're happy with the job then you don't need any more money'. We felt we were being cheated and eventually four of us went to see Mr Hasted together on a Friday night and asked for a raise. When we got the expected usual response, we all said 'No' and resigned our jobs immediately. It was our protest to him because we had all been offered other jobs and, by leaving straight away, it meant he would be short- staffed for the busy Saturday shopping day.

My next job was in a pub called The Coopers Arms in Chadwell Heath working on Saturdays. My role was to clean up and re-stock ready for the busiest day of the week, and then serve behind the bar from opening time. I learnt how to be a barman by working in the special function room of the pub that was let out for wedding receptions. It usually entailed just serving the drinks and ringing up the relevant amount on the till with the bill being settled by the bride's father at the end of the evening. The other barman was on the fiddle and used to ring up inflated amounts and then treat himself to a bottle of spirits to take home. He wasn't the only one to be helping himself because, one morning, I found a full bottle of whisky next to the dustbins. I took it back into the pub and waited to see what would happen. A bit later, one of the other cleaners asked who had been near the bins and when I admitted I had emptied some rubbish, he sheepishly asked if I had found anything!

I had my first holiday without parents when I was 16 and joined a couple of school friends, Graham Young and Len Merrifield, on a youth hostelling holiday in North Wales. After travelling by coach to Llangollen, the three of us typically journeyed 15 miles a day between hostels and had a great time walking in the summer sunshine until one morning we awoke to pouring rain. All of our hostels had been pre-booked so could not stay where we were. We started walking and were soon soaking wet even though we were properly kitted out with correct clothing. We decided to try hitching a lift but realised that we were unlikely to be offered a lift if all travelling together. We split up and put our thumbs out. I hadn't been hitching for long when a large American car stopped next to me and without further ado, I jumped in. It was only then that I realised that they hadn't stopped to give me a lift but to ask for directions to Mt Snowden! I quickly advised them that they would be unable to see the mountain as the weather was so appalling and recommended that they went to see Swallow Falls instead (coincidentally, very close to where I was going!). They accepted my recommendation and I guided them accordingly. I was the lucky one and had warmed up with a nice hot bath long before my friends arrived having had to walk the whole way.

Another day's trekking entailed climbing over some steep hills and, although the weather at low level was good, further up it was damp and very misty. Rather than risk getting lost due to the poor visibility, we decided not to continue over the hills but take a longer track at a lower level. Len and I took a recognised path down the hill but Graham decided to take a short cut. The grass was slippery and we suddenly became aware of Graham running down the hill quite out of control (Graham later told us that he'd slipped and once moving, couldn't stop). Len and I laughed so much at the sight of a cavorting Graham until he reached a rocky area when we saw him doing a somersault and landing on the back of his head. Luckily, he was only concussed and could still walk so we helped him back to the hostel we'd left a short while earlier for some first aid. Graham wasn't badly hurt and we continued our holiday after a good night's rest.

I was seventeen and walking home from the pub one day when I saw a Ford Prefect for sale in a local street for just £15. I didn't have a driver's licence at the time but intended to get one and thought it would be cheaper to learn to drive in my own car rather than pay for lots of lessons with an instructor. I returned with Dad later that evening and bought my first car. The only problem was that it didn't have any glass in the driver's door. That shouldn't have been a problem because Ford Prefects were very common cars and there were lots of 'breakers' yards' that sold second hand car parts. A replacement window should have been easy to find but Dad and I had no success when we scoured the local breakers' yard for one. Feeling a bit dejected, we were making our way back home after a whole weekend of searching when we saw a similar car for sale on a dealer's forecourt for £45. We stopped to have a look at it and as Dad had really enjoyed driving my car around (as he didn't have one at the time), he offered to make up the difference in cost. We negotiated the part-exchange of my car for £30 (£15 profit is eight days was pretty good going!) together with changing the wheels because the tyres were better. We became the joint owners of a very desirable car because it had a registration number plate of UGC 1. I had most use of the car because, once I had passed my driving test, I used to drive to school every day and even parked it in the teacher's car park as there were spare spaces. I was the only student in the whole school who owned a car and the first ever to be allowed to leave it in the car park!

The pub job was okay but it tended to interfere with my social life in the evenings so I found myself a Saturday job in a bookmakers (betting shop). When I was younger I used to watch horse racing with my grandfather, Frank Sadler, on TV during his visits or vice versa. We used to bet on the races using matchsticks and I usually won, much to his annoyance as he used to study 'form' whereas I just picked horses because I liked their names or I knew the jockey was good. I used to study 'form' when I was working in the betting shop because it gave me something to do first thing in the mornings when the shop was quiet. However, it was to no avail as I usually lost money on the day. (I should have kept to my old methods!). When I arrived on a Saturday morning, there was usually just the owner and me in the shop. I would be there to take the early bets, whilst the owner worked out the winnings of the previous evening's greyhound races.

On one particular day, there had been no customers before about 10.30am and I had finished my 'form' studying, had chosen my horses for the day and completed my betting slip, the first of the day. After the last race, I collected the outstanding betting slips because I had been successful in a couple of races and wanted to work out my winnings. Imagine my surprise when I found that my betting slip was not the first one in the pile (as it should have been). There were four other betting slips and they were all big winners! I suspected that the betting shop owner was stamping blank betting slips when he arrived in the mornings and then later filling in the names of winning horses. However, I couldn't initially understand why he would want to do that. I eventually worked out that he was defrauding the tax man as there was no tax to pay on race winnings but there was on shop profits! I decided not to say anything for fear of losing my job. There seemed to be lots of people 'on the fiddle' at that time and the bookmakers was no different.

When I'd joined the 6th form at school, I'd chosen Maths, Applied Maths, Physics and Technical Drawing as 'A' level GCEs subjects as these were my best subjects at 'O' level. At the start of my second year in the 6th form, I still had no idea what I wanted do after I finished school and the careers master couldn't help because he only knew about teaching! I thumbed through a university entrance book to see what was available and was somewhat disturbed to see that the vast majority of universities required a pass in a foreign language at 'O' level. That had not been explained to us when choosing our options at the age of 14 (although, even if I'd known it at that stage, I'd probably still have dropped French!) I decided to apply to Northampton College of Advance Technology (this became City University in 1966) in London for a course called Instrument & Control Engineering (because it sounded interesting), Leicester University for Physics and Pure Maths, and Salford College for Electronics. I wanted to have Leicester University as my first choice but my physics master persuaded me against it because he didn't think I would get a 'C' grade at 'A' level, which was their requirement. It is a sobering thought that my whole career and subsequent life changed at that point because, without my teacher's advice, I would have gone to Leicester as I actually achieved a grade 'B' in physics.

My 'A' level grades were good enough for me to be awarded the school 'Peake' prize for 'All Round Ability'. As soon as results were published, Northampton College called me wanting to know if I was going to accept the place they had offered me. Unfortunately, I hadn't managed to get sponsorship for the course (it was a five year 'thin' sandwich course whereby each year was spent with six months in industry and six months at college) even though I'd had interviews (ultimately unsuccessful) with British Steel Corporation, Central Electricity Generating Board and Plessey. Northampton College arranged an interview for me at a small electronics company called Dawe Instruments in Acton, West London and I was accepted for a student apprenticeship starting in September.

APPRENTICESHIP

School friends, Graham Turner, Jim Dumsday and I arranged a memorable horse-drawn caravan holiday in Southern Ireland during the summer of 1965. We travelled by train to Fishguard in Wales before taking an overnight ferry to Cork. I was never a good sailor and the 10 hour crossing in heavy seas did nothing for my stomach. Luckily, it was a warm night and the three of us stayed out on deck for the whole journey. Those who stayed inside had to put up with the horrendous smell of partially- digested and regurgitated meals! We collected our horse and caravan in Cork and were given appropriate instructions about caring for the horse and how much to pay farmers for allowing it to graze in their fields overnight. At our very first stop, the local farmer wanted double what we'd been advised to pay but we successfully bartered the price down. The farmer, however, decided to have a bit of fun with us and told us to put the horse in one field and our caravan in another. We'd only just settled down to our first home-cooked meal when we became aware of the farmer letting a herd of cows into the same field as our caravan. Not only that, but there was a bull with them! We weren't too bothered about not being able to leave the caravan as we all needed a good night's sleep after our ferry crossing so we turned in early for the night. We'd all been asleep for a couple of hours before being awoken by the caravan being buffeted by the bull attempting to gain access to two bags of oats that were strung in a rope cradle beneath the caravan. We shone torches out of the windows and that scared the bull away but he soon came back so we drew lots to see who should venture outside to retrieve the oats. Jim lost but still refused to go outside, and neither Graham nor I was going to volunteer. The bull eventually extracted one of the bags of oats and ripped it open for the herd to enjoy and we then managed a fitful night's sleep.

A few days later, whilst travelling from Bantry to Ballydehob, our horse threw a shoe. Jim and I went back to Bantry in search of a blacksmith. All this meant that by the time we reached Ballydehob it was much later than anticipated and we were too tired to cook for ourselves. Instead, we went into the local town in search of a meal and tried one of the very many pubs in the high street. There were no menus so we asked the barmaid if they served food. A fried meal was offered and we gratefully accepted and, about 20 minutes later, we were shown into the family's dining room to be confronted by a huge pile of food, much more than we could ever manage and all for a very small price. Towards the end of the meal, an elderly gentleman (the barmaid's father) joined us for a chat and asked if we'd like to play cards. Once we'd agreed, he asked if we'd like to play for money! I was willing but Graham certainly wasn't and so we just played for fun, which was just as well, as the old gentleman's rules were very different from ours. We eventually made our way back to the caravan which was in a field a short distance from the town. The sky was clear, there was no moon and there was no light pollution and the heavens were awe inspiring. There were so many visible stars that illuminated our path that we could see our caravan in the field without switching on our torches!

The final day of our holiday found us making our way back to Cork. Our female horse was normally very slow moving but when another caravan passed us pulled by a stallion, our horse pricked up her ears and set off in pursuit. Jim had been walking alongside our caravan but now needed to run to keep up and, because we were making such good progress, Graham didn't want to stop to allow Jim to climb aboard. Jim attempted to jump on to the front seat but slipped and ended up on the road and the caravan wheels went over him. The horse bolted and it ran several hundred metres before it would stop. I ran back to where Jim was still lying in the road and being attended to by some other road users. Luckily, the wheels had gone straight over his chest and, although badly bruised, he had no other injuries. He'd been extremely lucky! As I said earlier, a memorable holiday with several other amusing incidents that don't warrant mention here but will long live in my memory.

At the beginning of September 1965, I found myself some lodgings in Ealing, West London, which consisted of a small bedroom and shared bathroom. I had meals with the family during the week and went home most weekends. Dawe Instruments had taken on undergraduates mainly because of the government grants that were available at that time but didn't have any sort of training programme organised. I spent the first 3 months assembling printed circuits boards and the second half of the training period in a 'machine shop' learning to operate lathes and milling machines. I was operating one such machine when I realised that the piece of metal I was milling wasn't properly positioned in the vice. I wound the vice assembly away from the milling tool, loosened the jaws holding the metal and attempted to re-position it. However, the metal was covered in oil and extremely slippery and my fingers slipped into the still rotating mill tool that I had failed to turn off. There was blood everywhere and I was immediately rushed to the local hospital for treatment. Luckily, I had only cut the middle finger of my right hand but the whole nail was missing. The wound was cleaned and I was patched up quite quickly and returned to work where the news had spread around the whole machine shop. Another employee had previously lost the top of a finger and he gave me all the 'tales of woe' that he had experienced and what I could expect; not what I wanted to hear at that point in time. However, once the wound had healed, the nail re-grew and I was left with no sign of the accident. It was all my fault, as I knew I should have turned the machine off before making any adjustments, but 'health and safety' was paramount and an enquiry was launched. The result was that guards were fitted on to all of the milling machines that would prevent the occurrence of a similar incident. This made operation of the machines more difficult and I got a lot of 'stick' from the other machine operators for being so stupid at leaving the machine running whilst making an adjustment.

By this time, I'd sold my car and replaced it with a Lambretta scooter. However, the scooter didn't last long because it was too dangerous! I was returning to Chadwell Heath from Acton on a Friday afternoon when a police constable stepped into the road to stop the traffic and let school children cross. The road was wet and, as I tried to stop, the scooter skidded and I did a summersault in the air before landing in the road on the back of my head! Fortunately, as it had been raining, I was wearing a crash helmet (they weren't compulsory in 1965) and did not hurt myself. The police constable insisted that I must have been speeding but that was not the case; when he stepped into the road he didn't allow me enough time to safely stop in the wet conditions. Anyway, although the scooter was a bit scratched, it was still driveable and I continued my journey home. However, I decided that two wheels were not enough anymore and I bought a Ford Thames van which kept me mobile until I finished my 6 months in Acton. I kept the scooter to be used whilst at college in London for a few months until it was stolen.

In February 1966, I started my course at Northampton College and lived in Northampton Hall, Bunhill Row (the student halls of residence). I had a room on the 13th floor with magnificent views towards North London. There were 20 students on my course with about half of them living in Northampton Hall. The first semester at college wasn't too demanding and I passed all the exams at the end of July.

My second industrial training period found me in the 'Test' department of Dawe Instruments where, as the name suggests, we tested all instruments before they left the factory. One of the test engineers was a practical joker whose 'party trick' was to charge himself up with static electricity and then offer to shake hands with anyone who was passing by. Both participants received quite large shocks but this engineer didn't seem to care too much as he delighted in seeing the reaction of his victim! One of the pieces of equipment we used whilst tested was an ultrasonic cleaning tank which was filled with carbon tetrachloride, a very efficient cleaning fluid. Such cleaning fluids have to be handled very carefully now, as they are carcinogenic but this wasn't known at the time and everyone in the test department was exposed to the fumes. One of the favourite lunch time activities was to clean greasy car parts in the tanks and so the carburettor on my car was always sparklingly clean!

Martin Wightman was another student apprentice at Dawe Instruments who was also on the same course as me at college. We found some lodgings in Ealing where we shared a room and again had meals with the family. However, in this house, we weren't permitted to use the bathroom (that was reserved for the family) and we had to go to a communal bathing facility further down the road.

I'd played table tennis whilst at school and was one of the better players in my class so, whilst living in Ealing, I joined a table tennis club and my game improved enormously. Playing table tennis became one of my main leisure activities if future years.

Upon my return to college for my second semester, I found that Northampton College had received its Royal Charter and had become City University. I also found that seven of my student colleagues failed some of their exams, and the re-sits, and thus weren't permitted to continue on the course. I managed to get a place in Northampton Hall again and, as there were no exams at the end of this semester, I got elected on to the Hall committee as the 'Entertainments Member'. My main role was to organise the discos, dances and darts tournament etc. This role was particularly useful to me as I became the main contact for other colleges who wanted our students to attend their functions and parties! There was a lot of rivalry between the colleges in London and most of them had a mascot that was a target for other colleges to try and steal and hold to ransom. City University's mascot was a 5ft concrete carrot that stood in a locked frame in the university common room and it was widely believed to be so heavy that it couldn't be stolen. It was whilst preparing a disco in Northampton Hall that I and a few friends happened to be in the basement of the university to collect bunting and lighting when we came across our carrot. It had been hidden away whilst its frame was being modified following an unsuccessful attempt at stealing it. Anyway, we thought it would be a good laugh to move it to another part of the basement and anonymously call our President and claim to have stolen it. We found a caretaker's trolley and moved the carrot to a gents toilet cubicle, put it inside and locked the door. When we were back at our halls of residence, one of our group called the President and claimed to be from another college and demanded a cash ransom for the return of the mascot. The President agreed and said the ransom would be handed over during the disco that was being held that very evening. A couple of hours later, and after the President had confirmed that the carrot wasn't where it had been stored, he called me to tell me what had happened. He wanted my help to catch the carrot thieves when the ransom was being handed over so that they could be taken to the nearby fountains outside Britannic House, the headquarters of BP, and thrown in! Needles to say, I agreed to organise a gang to catch these thieves whilst making my little group aware of what was being planned. The President, whilst at the disco, announced his intention to pay the ransom but nobody came forward to receive it. The carrot was found the following Monday and stored more securely until its frame was reinstalled.

Martin and I used to play 'table football' in the Northampton Hall bar and became quite formidable opponents. The bar rule was that the winners of a game stayed at the table and the challengers paid for the next game. Martin and I managed to play many games without paying other than for the 1st game. However, we eventually found out that we weren't as good as we thought we were! We went to a pub with friends in central London that also had a table and we played a few games. Eventually, a couple of Italian lads came over and challenged us to a game. We were taught a real lesson on how to play and lost three straight games 9-0, 8-1, 9-0. I also played table tennis and darts and became university champion of both during my time at City.

My third industrial training period found me in the Drawing Office at Dawe Instruments and I found that particularly boring and managed to get myself transferred into the instrument development department, which was more interesting, but I got the distinct impression that I was being used as cheap labour rather than receiving meaningful training. Martin and I shared a bedsit in Ealing with a small communal kitchen. We had a gas fire in our room and our own gas meter that we had to keep 'feeding' with coins because we didn't get much gas for our money. However, we noticed that the padlock on the meter was identical to those we had at university for our storage cupboards and so tried our keys to see if they fitted. They didn't. That evening I went out to my table tennis club and, on return, Martin triumphantly showed me the unlocked meter and the adjustments he had made to give us more gas for our money. Martin had filed away part of his key so that it fitted the padlock! I laughed and, before leaving the room to make a cup of tea, suggested that when the landlord found out he would call the police and Martin's fingerprints would be found on the internals of the meter. When I returned about 5 minutes later, the meter was in pieces and Martin was frantically wiping all the parts he might have touched. We did, however, leave the adjustment so that we got a reasonable amount of heat for our money!

When we returned to University we complained to our tutor, Ron Watts, that Dawe Instruments didn't have any structured training for us and he promised to ensure that, when we returned, there would be something in place for us. I decided not to live in the halls of residence for that year and found a shared house in Muswell Hill, North London. It was a large house run by an Irish landlady and I shared a room with a friend on my course, Brian Strong, The food was awful. Every evening meal comprised 50% mashed potato, 20% cabbage, 20% carrots or peas and 10% meat except when she received a meat consignment from her husband, who was a farmer in Ireland, and then we got 30% very fatty pork (probably what he couldn't sell wholesale). One evening we returned from lectures to find that one of the two light bulbs in our room was missing. I queried this with the landlady and she told me that when she had come to our room the previous evening, both lights had been on and she considered that was unnecessary and thus she had removed one. She was unmoved when I pointed out that the only study table in the room was too small for us to both use at the same time and that was why I had been reading a book whilst lying on my bed. Brian and I and several other students moved out at Easter after only living in the house for about six weeks. The landlady was not amused!

Brian, Martin Fielding and I then found a living/bedroom plus kitchen (toilet in the garden) just 100 metres from City University for the three of us to share for just £5/week. It was rubbish accommodation but we figured that we would have our lunch and dinner in the university refectory, use the baths in Northampton Hall of

Residence and just use the room for sleeping in. Our parents would have been horrified if they had known but, to us, it was cheap and convenient. We found out later that the husband of the owner, who lived on the ground floor, was in prison for armed robbery and the father of the family on the second floor was with him. I had a birthday whilst living there and couldn't understand why I didn't get any cards on the day. That was until I found them in the dustbin! I reckoned that the boy who lived on the ground floor had found them on the doormat and opened them to see if they contained any money (fortunately not). Like father, like son!

One sunny and warm weekend, a few of my university friends and I went to my parent's house in Chadwell Heath and we were playing an illegally copied Beatles tape of their White Album. Brenda wandered into the room where we were all singing along to one the songs, Bungalow Bill. Brenda wanted to know who we were listening to and, for a joke, we told her that it was a recording we'd made. Brenda believed us and we didn't let on who it was. Some months later, Brenda was at work when the office radio started playing Bungalow Bill and she proudly announced to her work colleagues that her brother and his friends had recorded it. I'm sure you can imagine the office laughter and banter at Brenda's gullibility and embarrassment... I'm not sure I would have survived the night if I'd still been living at home!

APPRENTICESHIP

I became a Manchester United supporter after the Munich air crash in 1958 and it was 10 years later that Manchester United won the European Cup for the 1st time by beating Benfica by 4 goals to 1 in the final. That was in the days of Bobby Charlton, Denis Law and George Best, 3 of the finest players to ever play in a United shirt! It was a fantastic game that Brian, Martin and I watched in a local pub together with lots of other supporters. The pub landlord was also a supporter and when closing time came, he merely locked the doors and allowed everyone who was still there to continue drinking 'on the house'. We left in the early hours of the morning and staggered back to our accommodation which, fortunately, was only a few hundred metres away. The next evening we took part in a university car rally even though we were still feeling the effects of previous night out. We finished the rally without any unfortunate incidents!

When we returned to Dawe Instruments for our fourth training period we found there was still no structured training arranged for us. I immediately called our tutor and he came to Acton for a meeting with us and the personnel department. The net result was that we should resign from Dawe Instruments and that our tutor would seek alternative sponsors for us. We were both offered places with BP Oil, BP Chemicals and ICI. I accepted the offer from BP Oil, whilst Martin accepted

BP Chemicals. However, although we now had different employers (albeit associated) we were both sent to the BP Chemical site at Salt End in Hull. Our exposure to what happens on a chemical site was perfect for the careers we both wanted but Hull, as a city, left a lot to be desired! My memory of it was a place where the sun seldom shone. Admittedly, we were there from September to February and a winter on the NE English coast is never going to be ideal. However, the pubs were good, my darts playing improved and there was Hull University and a female teachers training college near by...

My fifth and final industrial training period was spent at BP's Sunbury research centre when I lived in nearby lodgings. My landladies were a couple of spinster sisters who thoroughly enjoyed talking at me. That was enough for me to spend most of my evenings in the BP Sports and Social club, returning late evening after having consumed a few beers. I would sometimes wake up with a bit of a hangover at weekends and then eat my breakfast in the company of the resident miner bird who was especially noisy (not what I needed when I had a thick head). A few weeks of that was enough and my grandfather, Frank, who was now a widower and living in nearby Acton, wanted some company and offered cheap accommodation. I moved in with him for the remainder of the 6 month training period.

My project whilst at Sunbury, and then taken further

for my university final year dissertation, was to automate a testing procedure known as "Fraas breaking point test for bitumen". The Fraas breaking point test is one of very few tests that can be used to describe the behaviour of bitumen at very low temperatures. It is essentially a research tool that determines the temperature at which bitumen reaches a critical stiffness and cracks. In the Fraas test, a steel plaque is coated with a thin layer of bitumen and is slowly flexed and released whilst its temperature is lowered until the bitumen cracks. The equipment BP used for this test was somewhat time-consuming and didn't produce consistent results. A laboratory technician would try to reduce the temperature by dropping small pieces of solid carbon dioxide into a flask containing the flexing apparatus and, by sight, sense when the bitumen sample had cracked. I took the basic equipment and made modifications so that the flexing plaque was cooled at a constant rate by blowing refrigerated air over it and used a small microphone to detect the cracking of the bitumen sample.

I shared an office with an engineer called Bill who was in his early 30's and was a bit of a boffin type. He'd been recently married and was quite proud of his method of finding a suitable wife! He confessed to have never been 'in love' with a girl but nevertheless wanted to get married so he made a list of all the unmarried women he knew (some of whom he didn't really know) and gave them marks out of 10 for the attributes he considered important in a wife. These included good looks, ability to cook, good sense of humour, level of education, etc. He didn't always know about their cooking abilities so would start with a mark of 5 and then adjust it if he gained information from friends and work colleagues. After about six months of 'research', he totted up the scores and set about pursuing the woman with the highest score. Amazingly, this method worked for him and he was married two years later. I'm not sure what his wife thought about his method of selecting her but it's certainly a very unromantic way! BP had recently sent all staff engineers on a course to help them write reports in simple language that most people would be able to easily understand. Bill thought this course had been really useful to him and he set about using these newly-learnt techniques in his monthly progress report and was somewhat peeved when it was returned to him by his boss who thought Bill had now over simplified his language. Thereafter, Bill would write his monthly reports in simple language and the pair of us would then spend several hilarious hours converting it in to highly technical language before submitting both reports to his boss.

Living with my grandfather was not easy and although he wanted company, he also wanted to watch his favourite TV programs (and that was his right). I had no desire to sit and watch the sort of programs he liked (Coronation

Street); I had a good social life and spent a lot of evenings out. Frank didn't like the fact that I spent so much time away and complained to my mother. I tried to spend a bit more time with him but we didn't have a lot in common, and the TV was a constant irritation for me. The months soon passed and it was back to City University for further study and final exams.

Graham Turner, with whom I'd enjoyed the holiday in Ireland, also studied at City University but in the opposite six months until the final year when he was also moved into Northampton Hall in February 1970. He'd been a friend during my time at school but he'd a reputation for being 'economical with the truth' when relating some his experiences. Everyone knew this and just accepted it as being part of him. Graham asked if he could borrow my car, a Ford Anglia, one weekend to visit his sister and her boyfriend who were living in Southend. He'd borrowed it before so that wouldn't normally have been a problem but, this particular weekend, I had arranged for a local garage to replace the brake shoes that I knew were worn. Graham then offered the services of his sister's boyfriend, who was a mechanic, to do the car repair if I lent it to him. That seemed a good deal so off Graham went with my car. On the Sunday afternoon, I received a telephone call from Graham's sister telling me he had been involved in an accident, was in hospital and letting me know where the car had been towed to. Martin and I set off to the hospital to see Graham, who wasn't badly hurt but being kept under observation. He explained to me that the steering had failed and he went straight on at a bend and collided with a tree. I then went to see my car to establish the extent of the damage and, unfortunately, it was beyond repair. However, I tested the steering and it was OK! I went back to the hospital and met Graham's sister, Gill, who also confirmed that her boyfriend had checked the car after the accident and found that the steering was still working normally. Gill also told me she knew nothing about the brakes being repaired over the weekend. Once Graham had been released from hospital, we talked again about what had happened and he still insisted that the steering had failed and that the brakes hadn't been repaired because the parts weren't available. I knew that this was just a story he was making up so that he could justify, to himself, not paying anything towards my 'written off' car. That was the end of our friendship!

Bridge Over Troubled Water by Simon & Garfunkel was released in January 1970 and, whenever I was in the hall of residence bar during my final semester, which was frequently in the evenings after revising for final exams and then playing darts to unwind, this album was playing and became synonymous with the period. Now, whenever I hear one of the album tracks, I'm somehow transported back to that time and many happy memories around the people, places we visited, and events, including Brenda's marriage to Clive at St Andrew's church in Weeley. Clive had been an instant 'hit' with Brenda when she met him but the same cannot be said about my father's view! Clive was always one to speak his mind and this didn't go down well with Dad. However, his view quickly changed and Clive was welcomed into the family on 25th April.

It was on Thursday 30th April 1970, that I noticed a poster in the bar of Northampton Hall advertising a disco at Rachel Macmillan (teacher training) college in Deptford. It was an unusual night for a disco but I was bored with revising for final exams so I persuaded Geoff Jones and another student to go along with me. That was the night I met one Joan Seaborn and my life would never be the same again. However, I later found out that when Joan met up with her college friends after the disco, she told them I was a nice guy but not the type she'd marry! Even though I needed to revise for final exams, Joan and I met up most weekends to spend some time together. After exams had finished, everyone on my course invited our tutors to a celebration dinner at Northampton Hall, and I invited Joan to come, too. I picked her up from Deptford in my Mini (replacement for the Ford Anglia), which she suggested needed cleaning before we went for dinner. We spotted a drive-through car wash facility and duly drove in between the large rotating brushes. My Mini was a very early one that had sliding windows that weren't particularly well sealed and, once the brushes started turning, water was pouring into the car. Joan and I grabbed some dusters that happened to be in the car and tried to stem the flow of water but it was to no avail. Joan and I couldn't help but laugh at our predicament and huddled together in the centre of the car until our ordeal was over. We were both soaking wet and I had to take Joan back to her college to change before driving to Northampton Hall, also to change, before our evening celebration. The Mini was never treated to a car wash again and was sold soon after!

After final exams, Graham Burke, Jim Pollitt, Martin Wightman, Geoff Jones, Brian Strong and me, all from City University, took two cars, two tents and £50 each and headed towards Italy for as long as we could survive until the money ran out. We didn't get far into France before one car, a Hillman Imp, blew a head gasket. That cost us several days and quite a lot of money to fix. However, we were soon on our way again only to find that the garage hadn't done a very good job and the gasket blew again! This time we decided not to fix it but be careful to keep an eye on the temperature gauge and top up with water when required. We managed to get all the way to Viareggio in Italy and back to London without the engine failing us.

Once back home again, I got my degree results and was happy to learn that my 'upper second' was the highest awarded on my course and that I had won the 'Course Achievement' prize. As a present to myself, I bought a sports car, a Triumph Spitfire, from a dealer in North London and, without telling Joan what I'd done, went to London Heathrow airport to meet her from her holiday in Portugal with one of her college friends. Her delight in riding in a sports car was, however, short lived! On driving back through Holborn, in central London, the front suspension collapsed and the car was not driveable. The garage that collected my car told me that the trunnions (part of the suspension) were so badly worn that they had broken. That was quite a surprise to me as the garage that sold the car had recently issued a new MOT (Ministry of Transport) certificate indicating that the car was roadworthy. I complained to the dealer and threatened to report them for selling an un-roadworthy car if they didn't repair the car at no cost and he duly complied. There were a number of unscrupulous car dealers around at this time and he was obviously one of them.

SERIOUS WORK BEGINS

Now that my student days were over, BP asked where I would like to work. After five years of moving every six months, I wanted a bit of stability and, as Joan was going to be in London for at least another year, I opted for head office at BP House in Moorgate and started working there in September. My job was to design control systems to enable automatic control of oil refineries. Two other BP student apprentices who graduated with me, Graham Burke and Jim Pollitt, also opted for head office and we decided to rent a house together in Lee Green, SE London. I had only been working in London for about six weeks when my manager called me into his office and told me that I was to be transferred to Grangemouth in Scotland for two years to get some more site experience. That was not what I wanted to hear as I wasn't sure my relationship with Joan would stand the separation. However, as I had signed a six month lease agreement to stay in the rented house, BP decided to postpone my transfer until it expired. It was now decision time, move to Scotland or allow my romance to develop. It was probably not the best career decision I've ever made (however, I've had absolutely no regrets) but I decided I wanted to stay in London and so started looking for alternative employment. I quickly found a job with Automatic Control Engineering in Sidcup, Kent, and resigned my position with BP.

Joining Automatic Control Engineering was also not a great career decision as they were only a small company with very limited prospects but at least it enabled me to stay in the London area and close to Joan. My relationship with Joan had been growing stronger and, when she finished her teacher's training course, she also opted to stay in London. I had known Joan for about 18 months when I proposed to her and was 'over the moon' when she accepted. I formally asked Joan's uncle, John (Joan had been brought up by her aunt Freda and uncle after her parents died[1]), for permission to marry and we got engaged at a Queenswood (Joan's old school) ball in London. The wedding date was set for 5 August 1972.

About nine months after joining Automatic Control Engineering, the company was suffering from a lack of orders for their control panels and decided it needed to reduce its headcount. I was surplus to requirements and was made redundant. Now I had no income but still had rent to pay, a car to maintain, food to buy and a wedding to save for...

---

[1] Henry Charles Seaborn died from lung cancer when Joan was just 8 years old. His wife Mary (nee White) then suffered from deep grief over the loss of her husband until, tragically, she took her own life 2½ years later.

SERIOUS WORK BEGINS

I can't really remember how I found out that Blue Circle Cement wanted to recruit a junior Instrument Engineer but I was offered a position in their Southern Area Technical Services department based in Northfleet, Kent in January 1972 and duly joined them. Joan & I went 'house hunting' near to where I was working and we bought our first home at 63 Lyndhurst Way, Istead Rise, Kent for £9000 and I moved in during May of 1972. We had no furniture and I remember sitting in our lounge in a garden chair on an 'off cut' square of carpet watching TV on an old black and white set. We went to a few auctions and bought some second-hand furniture but we treated ourselves to a new bed!

MARRIED LIFE

On 5th August I awoke to the sun streaming in to the window of my hotel room for the start of a perfect day. My best man, Graham Burke, and I ate a hearty breakfast and set off for the church. Our wedding was held at All Saints parish church in Ladbroke, Warwickshire and I was the happiest man in the world when I saw Joan walking down the aisle to become my wife. Our reception was held at John & Freda's house, The Abbey, in nearby Southam. Their house had a large and lovely garden and a marquee was set up for the function. Much of the day now seems a blur but everybody had a wonderful time.

One amusing incident from the day was related to my car. (I now owned a Ford Cortina). I knew Martin had a key that would open the car door (lots of Ford cars had keys that would work in other Ford cars) and thus I laid a deception! I left the car in a prominent position just outside of The Abbey with an empty suitcase on the back seat. Needless to say, the car was opened and, upon finding our empty suitcase, our wedding guests assumed the car was a decoy and went looking elsewhere for a car they could decorate. When it was time for us to leave to go on honeymoon, we dashed to the Cortina (still not decorated) and locked ourselves in. However, the car wouldn't start as somebody had opened the bonnet and swapped over the spark plug leads. Eventually, and after the car had some last-minute decorations added, the leads were swapped back for us and we made our getaway and picked up our packed suitcases from a friend of Joan's who lived nearby. Off we went to Luton airport for our honeymoon flight to Majorca, Spain.

In retrospect, I should have arranged for us to stay

5th August 1972

overnight in an hotel in England but there had been a convenient early evening tour company flight to Majorca that should have meant we arrived at our hotel before night fall. Unfortunately, our flight was delayed and, once on board the aircraft, we were told that our hotel had been over- booked and we had been transferred to a sister hotel in Cala Millor, a three hour drive from the airport (instead of the selected one that was much closer). To cap it all, our minibus driver had spent the evening drinking whilst waiting for our delayed flight and I think we were fortunate to arrive at our destination hotel at 3am! The rest of our honeymoon was idyllic and the hotel was better than the one I had booked.

Upon our return to England, Joan moved into our house and applied for a teaching position in the local infants school, which she duly got. Joan had a five minute walk to school and I drove to Northfleet which took about ten minutes by car.

My work at Blue Circle Cement was interesting and I was involved in a number of projects in the South of England. A notable one was to detect 'blobbing' from the chimney stacks. Blobbing was a phenomenon where cement dust, which was supposed to be collected in a precipitator, was electro-statically charged and, as it inadvertently went up the chimney stack, the dust particles stuck together to form globules about 0.5cm in diameter. These globules would eventually settle on the surrounding area and the company would get complaints from local residents about dirty washing and cement being deposited on their cars. I developed an 'early warning' system that used radar to monitor the tops of the chimney stacks to see when globules were being discharged. We were somewhat surprised a few months later when local residents were again complaining about deposits on their cars even though they'd been assured by the company that we'd installed a detection system. Some of the residents called for a public meeting to try and get compensation and/or the plant shut down and arranged for samples of the 'pollution' to be tested to prove it came from our facility. That meeting never happened because, when the results of the tests were returned, they showed the deposits to be acidic whereas any possible pollution from our plant would have been alkaline. The local power station was to blame, not us!

Another of my functions was to be a commissioning engineer for starting up cement kilns and so I was sent to a Blue Circle plant at Snodland, Kent to gain experience in how to start up and control rotating kilns. I was assigned to work normal daytime hours of 9am to 5pm whereas the kiln burners (the men who operated the kilns) worked shifts: 6am to 2pm, 2pm to 10pm and 10pm to 6am. That meant that I saw the shift 'hand over' at 2pm and was somewhat surprised to see what was happening. The kiln I was assigned to was burning normally at 2pm when the next kiln burner arrived and he immediately changed some of the controls that cause the kiln to burn inefficiently for a couple of hours before it settled down again. After I'd witnessed this for a couple of days I asked why he changed the settings every time he came on shift and, with a knowing look on his face, informed that the morning kiln burner, Joe, always changed the settings just before his shift finished because Joe didn't want his rivals to know how he operated the kilns. Apparently, this was a little trick that the previous kiln burner did so as to hide the optimum control settings from his rivals. I kept a very close eye on what happened the next day and satisfied myself that there was no change of controls just prior to shift change but, no matter what I said, the next kiln burner would not believe me. All this meant that off-specification clinker (that's what the kilns produced in the cement-making process) was made for a couple of hours following a shift change and I was too junior to be listened to when I commented about what was going on.

I'd been a member of the City University Motor Club and enjoyed competing in a few rallies so the Blue Circle Motor Club was particularly attractive to me. I entered their rallies in my Mini with Graham Burke as my navigator. The skill was mostly associated with the navigation but, unfortunately, I couldn't do that because I became car sick if not watching the road. However, Graham was very good and we were quite successful and even won a Marlboro-sponsored night rally. The club also ran an annual car gymkhana that Joan and I entered together with Brenda and Clive as another team. The events included driving around a course as fast as possible with a cup of water on the bonnet (penalty seconds were added for spilt water), driving with the driver blindfolded and guided by commands from the co-driver. Joan and I had a successful and lucky day winning many prizes including 1st prize in the raffle (one of my nicknames was Jammy Roger). The main event of the day was to drive around a course on a playing field that included reversing into garages, handbrake turns, etc. and stopping astride the finishing line in the fastest time. Again, my luck was in as we had a shower of rain just before the event started that made the surface slippery and negated the normal advantage of cars with powerful engines. Many of the cars that went before me tried to go too fast and collided with course markers, which cost them penalty marks, and slid over the finishing line and had to reverse back before their time was assessed. When I was waiting at the start line, a fellow competitor approached me wanting to know if my secret was using coal as a fuel instead of petrol! This was because there was a lot of black smoke coming out of my exhaust. However, I had the last laugh as my knackered Mini engine didn't have the power to spin the wheels even on wet grass and I went round the course in the fastest time and Clive, who also borrowed my car, went round in the third fastest time.

Work wise, we were particularly busy on a whiting (paint colour additive) project at Swanscombe, Kent and recruited a contract instrument engineer, Les Holton, to assist with the design. Les had worked for a number of companies and, as a contract engineer, tended not to stay at any one place for too long and so left after about a year when the project finished.

A month or so later, Les called me and offered me a job with him at a newly-started company, Ameron, working in the Oil and Gas industry. The company was American-owned and well-established in the USA and they wanted to set up an office in the UK to get involved in the North Sea oil sector. Their problem was that they didn't need a large work force until they had been awarded a contract but wouldn't win any without at least some workers. The offer was 50% more than I was earning at Blue Circle Cement and the prospects were very enticing so I duly moved on. Ameron was based in Croydon and so I now had a 45 minute drive to the office.

This move to Ameron was also not great for my career! As Ameron had no project work, we were involved in bidding for some. However, there weren't too many projects to bid for (we only bid for two projects in the six months I was employed there) and so there was a lot of sitting around to be done but it did enable Les and I to become quite proficient at the Daily Telegraph crossword.

When we started doing the crosswords, Les and I would spend nearly the whole day on them and only manage to solve about 50% of the clues. At the end of each day, the office experts would explain our unsolved clues. After six months, Les and I would have our own copies of the crossword and complete them in a couple of hours! The boredom eventually got the better of both Les and I and we started looking for other opportunities. I joined an agency, Pipco, who specialised in providing contract workers in the petrochemical industry and they soon found a potential position with Monsanto, an American chemical company. The interview went well and I was offered a position as an Instrument Engineer in their London office, which was next to Victoria station. The remuneration was now double what I had been earning at Blue Circle Cement but I had to cope with a daily commute to London.

MONSANTO

The London office of Monsanto (Central Engineering Division) was effectively a design contractor but only involved with projects for their own sites, of which there were 15 around Europe. I was responsible for the design of control systems for their chemical plants. I hadn't been there long when there was a fire on one of their sites (operated by Forth Chemicals within BP's site in Baglan Bay, South Wales). I was duly sent to help with supervising a local work force to get the plant producing once again. I must have impressed the construction manager as, at the end of the reconstruction period, he wrote a report which detailed what had been achieved and recommended that I be offered a permanent staff position within the company. My manager, Vic Verco, made an initial approach to me but couldn't offer me a salary that was close enough to matching my contract rate.

Upon my return to London, a new project (a plant to produce a new agricultural chemical) was planned for the Antwerp site which required using a design contractor, Crawford & Russell, based in The Hague, Holland. I was chosen as the instrument engineer and was now involved in travelling to Holland every week to monitor the design work. It was a 'high profile' project within Monsanto and they sent a project manager and project engineer from head office in St Louis to lead the project team. I'd been working on the project for quite some time before the project manager discovered that I was a contract engineer. This caused him some concern as contract engineers were only on a short notice period and could thus leave within a week. He wanted to know why I worked as a contract engineer rather than permanent staff and I told him it was purely a matter of what I was earning. I also assured him that I had no intention of leaving before his project had been completed. The American project manager obviously put pressure on the Human Resources department and I was offered a generous permanent staff salary. I duly accepted!

Commuting to London and frequent flights to Europe meant that Istead Rise was not an ideal location and so, in 1975, Joan and I decided to look for a new place to live that had better train access to London and was closer to an airport. Copthorne was a nice village close to Gatwick airport and the main London to Brighton railway line and we soon found our new home, 11 Newlands Park, Copthorne, West Sussex. Newlands Park was a development of about 30 detached houses all situated in quite large gardens in a park like setting.

Copthorne had an active Residents Association which I joined as a way of getting to know the area and people in the village. I hadn't been on the committee very long when the chairman received a letter from the council seeking Residents Association input into a planning strategy document they were preparing. The council wanted to know where the Residents Association would prefer new housing to be built. The general consensus amongst the committee was that the village was large enough already and that the local infrastructure couldn't support more residents. I tried to argue that this approach would not stop more housing being built in Copthorne and that we would lose the opportunity to have any influence on the position of it. I was a lone voice and the committee eventually prepared a response to the council titled "No More Development, None!". Needless to say, there were more houses built in the village and the residents had to accept them wherever the council thought best. I still find it difficult to understand the 'head in the sand' attitude of some people who can't or won't accept that change is inevitable and that it's better to try and influence those changes rather than just reject them.

Joan was pregnant when we moved and so didn't continue working as a primary school teacher. We were obviously very excited about becoming a family but, towards the end of Joan's pregnancy, she was becoming uncomfortable so we decided to try and help things along! There was a hump-backed bridge on the A22 road not far from where we lived and we thought a few trips over it might help bring on labour. It didn't work and so we just had to wait and let nature take its course. On 28th February 1976, in Crawley hospital, our beautiful daughter, Karen, was born and so our lives changed forever!

Looking back on that time, I now realise that Karen was a relatively easy baby for us to adjust to. She rarely cried, and when she did, it wasn't for long and her sleep patterns weren't as bad as those some other local parents were experiencing. I think I got off lightly because Joan respected the fact that I was the one working and so tried not to wake me in the night too often. The early morning rousing meant that I just went to work earlier and then caught a correspondingly early train home again in the afternoon. The old adage of 'early to bed, early to rise' became the norm for us for a while. Also, our carefree lives of doing what we wanted, when we wanted, were substituted with endless rounds of nappy-changing, hanging washing on the line, etc. but that is something all parents are familiar with. However, those negatives were more than compensated for by all of the joy we felt watching our daughter growing up and experiencing the world around her.

Later in 1976, the chemical plant in Antwerp was being built and I was part of the team sent out to monitor the final installation and commission of it. This was going to take about 3 months and, although I could have done a weekly commute to Antwerp, Joan and I decided to rent a local apartment. It now seems difficult to believe but we put a roof rack on our MGB GT and packed everything we needed for ourselves and baby Karen into our car for the duration of our trip. Our apartment was on the 2nd floor of a 3 storey block in Deken Jozef Lensstraat. The commissioning instrument engineer, Bob Howie, lived on the ground floor and his wife, Helen, who became a good friend to Joan whilst coping with the pressures associated with caring for a young child in a foreign country.

Helen also had a young son and so was grateful to have someone around to share time and 'baby-sitting' duties with. One of our first concerns in Antwerp was associated with mosquitoes! They were such a local problem that there was even a bye law forbidding residents from leaving water in bowls for their pets in case they became breeding grounds. Joan and I could cope quite easily but we didn't want Karen to suffer from bites so our ingenious solution was to relocate the apartment's net curtains and use them to cover Karen's cot. Luckily, Karen was still so young that she didn't stand up and thus couldn't become entangled in the curtains.

Bob and I would sometimes travel to the Monsanto site in each other's cars and this arrangement worked well but only for the first couple of weeks. Unfortunately, Bob liked a drink or two after work and there was a local bar, called Peyton Place, in Lillo Fort, close to where we were working. Whenever Bob was driving, we'd call in for a drink where we'd meet up with contractors who were also working on the site. Drinks were usually free to Monsanto employees, courtesy of the contractor bosses, because of the money they were earning from Monsanto. It was not unusual to get quite a few beers lined up at the bar because whenever Bob's glass was nearly empty, someone would order another round of drinks (you weren't asked if you'd like another beer). I had no desire (or capacity!) to drink as many as Bob so whenever a new drink arrived for me, I'd leave my current half-finished glass elsewhere in the bar before collecting my new one. In the end it was better to travel to work separately so I could avoid Bob's ritual drinking sprees. It also became traditional for there to be a 'send off' drink at Peyton Place for everyone who had finished their involvement and was leaving the project.

Alvaro Canesares was a young process engineer who had been seconded to the project from our head office in St Louis and, on his last day, he invited everyone for a drink. I went along but didn't stay very long before going back to my apartment. The following morning, our project manager called everyone together and asked if anybody knew where Alvaro was because he hadn't been seen since leaving the bar and hadn't returned to the hotel he was staying in. Nobody had seen him and the story then unfolded that he'd left Peyton Place having drunk far too much and drove back towards his hotel in foggy conditions. On approaching a set of traffic lights, he failed to stop in time and collided with another car. When he got out of his car, he offered to pay for the damage but the other driver realised he was drunk and said he going to call the police. Alvaro panicked and set off again in his car but took a wrong turning which took him towards the docks area. Unfortunately for Alvaro, there was a police patrol car in the area which set off in hot pursuit. Whilst driving too fast, Alvaro came across a goods train that had stopped across the road (there were no gates) and he couldn't stop in time and thus collided with it. Alvaro now left his car, a few belongings, including his passport and, because he knew the police were close behind, ran off across some waste land adjacent to the docks area. We were all naturally concerned that Alvaro could easily have fallen and broken a leg whilst on the waste land so we organised several search parties to go and look for him. Having had no success, we all eventually returned to the Monsanto site to learn that Alvaro had been picked up by the docks police the previous evening and spent the night in the cells. In the morning, the traffic and docks police liaised and eventually informed the Monsanto site manager about Alvaro's whereabouts. Alvaro spend the next 30 days in prison awaiting trial and, upon being found guilty of several offences, was deported back to America. In partial mitigation, his defence lawyer had explained that Alvaro was of Columbian descent and running away from police, rather than stopping, was a natural response when in trouble! I'm not sure if that would have impressed the judge...

Monsanto was a great company to work for and they promoted employees to responsible positions at relatively young ages. Such was the level of responsibility given to me that I became a Chartered Engineer with the Institution of Electrical Engineers in 1976 at the age of only 29 years, relatively young to be granted this status. I was also appointed as Monsanto's European representative on their international Process Control Steering Group, making annual visits to the USA for meetings to assess and recommend control systems and instrumentation for their plants around the world. This was quite a perk as I could often extend my trips by a couple of days to see some of the local tourist sites.

My first visit to America was for a conference in Houston, Texas where I stayed in a hotel close to the venue. I arrived late in the afternoon, having taken a taxi from the airport, and managed to stay awake until about 8pm local time (2am UK time). I awoke on a sunny Sunday morning at about 4am, dozed until 5.30am and then got up very hungry and went in search of some food. Unfortunately, this hotel didn't start serving breakfast until 7.30am and I couldn't wait that long. The hotel receptionist advised me that there was pancake house that would be open about 1 mile away and, as I didn't have a car, set off walking along the side of the road which was very quiet at that time of the morning. I was about half way to the pancake house when I saw a police patrol car cruise by on the other side of the road and was somewhat surprised a few minutes later when he stopped his car beside me wanting to know where I was going. I explained about my hunger and lack of breakfast facilities at the hotel which didn't impress him and I got a lecture about the dangers of walking along the highway and the possibility of being mugged. He knew from my accent that I wasn't a local and he took pity on me. He offered me a lift to the pancake house and, as I got out of his patrol car and thanked him, he told me he would pick me up again in an hour and take me back to the hotel. What service!

The next project at Monsanto was to design and build an agricultural research facility (known as growth rooms) at Louvain-la-Neuve, Belgium. We didn't know it at the time, but Monsanto's head office in St Louis was originally tasked with responsibility for the design but they didn't think they could meet the design criteria of temperature 5 to 30 +/- 0.2 oC and 50 to 95 +/- 2% RH (Relative humidity) within rooms, so sent the project to London. Fläkt, a Dutch air conditioning company, was appointed to design and build the facility but, very early on, we realised that they lacked personnel with sufficient experience to complete an adequate design. The decision was taken to bring the design responsibility back to our own offices but to retain Fläkt to actually construct the rooms. At the end of the detailed design, our lead engineers moved to Belgium for the installation and commissioning phase of the project. Although Louvain-la-Neuve is some 20km East of Brussels, we decided to go back to the same apartment block we'd previously rented in Antwerp, together with Alan Drake and his girlfriend Jayne, and Sid Matthews and his family. We men would travel about 50km each way by car to the site, whilst the ladies had a great time socialising together.

The growth rooms were commissioned early in 1978 and the local researchers were delighted that we not only met, but exceeded the design specification. The project manager later congratulated the whole team but singled out Alan Drake, for the heating and ventilation design, Peter Gonnet for the lighting design (that had to simulate natural light) and me, for the control systems and for being 'the quiet one' working behind the scenes making sure everything 'hung together'. The description of me as 'the quiet one' probably typified my role in projects throughout my career. I had no ambition to become a project manager because I enjoyed the technical aspects of projects rather than ensuring they were on schedule and within budget, but always wanted to ensure their success. Thus, I would often delve into the technical side of other engineering disciplines to make sure everything was 'hanging together'.

My father died on 27th April 1978 and, naturally, this was a very sad day for me. A couple of years earlier, my father had been coughing in the bathroom and my mother noticed that he'd spat out some blood. It later transpired that this had been happening for some time but Dad wasn't one for visiting doctors; now, however, my mother knew what was happening and had an appointment arranged for him. My father had been a smoker of cigarettes since he was a teenager but had stopped some ten years earlier. Unfortunately, the damage had already been done and he now had lung cancer. He had an operation to remove the tumour at the Royal Brompton hospital but, when his chest was opened up, it was discovered that the cancer had already spread and there was little the surgeons could do for him. It was early in the afternoon on the day that he died that I was sitting beside his bed at home when his breathing changed and I heard what is commonly known as the 'death rattle'. Although I hadn't heard it before, I knew what it was but didn't know how long it would last. As it happened, it was only for a few minutes before he slipped peacefully away and I called my mother, who was in an adjacent room with Joan. Karen was at Brenda and Clive's home on that day and, when she visited the following morning, she ran into the bedroom, stopped, and I still remember her exclaiming "Grandad's gone" (A lump still comes into my throat whenever I think about that incident). Grandad was cremated at Weeley crematorium and his ashes were interred in plant pots in his garden at Weeley Heath, Essex.

I was the executor of my father's will and I was shocked to find out how little he had been earning as an office manager at Allen & Hanburys (later acquired by GlaxoSmithKline). He had been responsible for their payroll department for over 20 years and, although he received 'cost of living' increments each year, he appeared to have had little else in the way of increased pay. I realised that my parents had needed to be careful with their money whilst I was growing up, for there was very little spare.

26th July 1978 was another memorable date when our daughter, Michelle, arrived on the scene. Had Michelle been our first child then she would probably have been an only child! Michelle didn't sleep through the night for over two years. I was lucky that I travelled a lot with my job and so could get a good night's sleep when staying in hotels. However, when I was back for weekends, Michelle was definitely my responsibility during the night.

APRIL COTTAGE

It had been clear that we would need a bigger house and, although we considered extending our Newlands Park home, Joan and I decided to move. We signed up with several estate agents but, as Joan and I wanted to stay in Copthorne, the number of potential properties coming on to the market was very few. A letter containing details of a potential home arrived one Saturday morning but, when I called the estate agent to arrange a viewing, I was told the house was already 'under offer'. I wanted to know how that could be the case when I'd only just received the details but I didn't get a satisfactory answer. I later found out that this agency kept a 'card filing system' (there were no computers in estate agency offices in 1978) to keep details of what prospective buyers wanted and, when a new property came on to the market, they started at the front of their alphabetical index and sent out the details to the first few potential buyers. If the property was unsold a week later, they'd then send details to the next few in the index. With a surname beginning with S, we'd always be late in receiving details of potential properties. We eventually bypassed the estate agents and delivered letters expressing interest to buy if/when available to fifty houses in the village that looked, from the outside, as if they might meet our requirement of a character home, 4 bedrooms and a large garden. To our astonishment, we received seven positive responses and, after visiting them all, reached an agreement to buy April Cottage, Copthorne Common, a five bedroom, mock Tudor house (built in 1953 as a copy of a real Tudor house near Gatwick airport), ½ acre of garden with a stream running through it. We moved in during the spring of 1979.

April Cottage

It was at about this time that I joined Crawley Bridge Club. I first started playing bridge when I was in the 6th form at school but the game we played was more like 13 card 'brag' than bridge. I'd continued at City University and even represented them in an inter-university competition. There were a few bridge players at Monsanto and we used to play a few hands at lunch time. Mike Clarke was one of my partners and he lived not too far from Crawley and so we both joined the club at the same time. My ability as a bridge player gradually improved and I eventually represented Sussex in inter-county games.

Monsanto owned Fisher Controls, an American control systems manufacturer, and we had to use their equipment wherever they could offer a suitable product. In 1979, Fisher Controls launched a microprocessor control system called Marcus 16, one of the first such control systems on the market. It was a programmable system that had a total memory of just 32 kilobytes; about half for the operating system and the rest that could be programmed for control purposes. We had a project at our Ruabon site to automate a chemical batch process and thus we were tasked to use an early version. It turned out to be somewhat under-powered for what we needed but we managed to programme it for the required functions and we successfully started up the plant in February of 1980. Joan, Karen and Michelle came up for a few days holiday in Llangollen whilst I was involved in the start-up and we decided to show Karen where Daddy was working. We drove up to the site fence and Karen could see the multitude of pipes, tanks, valves, etc. that comprise a chemical plant. Karen was clearly puzzled and eventually looked up at me and said "I can't see the flower!" Joan and I were in hysterics but, of course, it was entirely logically to a 3 year old that if Daddy went off to the plant every day, you'd expect to see flowers!

The newly-automated plant was successfully commissioned and the design team went back to London but, after only a few weeks, the control system 'tripped' and the batch process stopped. This wasn't serious and the plant operators merely restarted the process but raised the query as to why this had happened. We had no idea and didn't spend too much time investigating the problem until it re-occurred about a month later; thereafter, the frequency of 'tripping' increased until just before Christmas it became intolerable. Fisher Controls were called in with their diagnostic equipment to see it there was any pattern as to when the trips occurred, but none was found. There was a lot of scratching of heads going on and I eventually contacted the CEGB (Central Electricity Generating Board) because I knew they'd also installed a Marcus 16 and I wanted to know if they'd experienced similar problems. They'd started up their system about 9 months previously and had no problems at all. I now requested a full list of companies that had installed Marcus 16s and, although quite a few had been sold, not many had been commissioned. However, there was one that had been commissioned at a carbon black plant in Glasgow and I duly contacted them for their experience. Imagine my surprise when told they'd had identical problems that had been sorted out by Fisher Controls about 3 months earlier. It appeared to be a case, with Fisher Controls, of 'the left hand not knowing what the right hand was doing'. Anyway, we now knew what had caused the problem and could remedy it. The Marcus 16 was an early microprocessor control system that held its programme on an EEPROM (Electrically Erasable Programmable Read-Only Memory). Our Marcus 16 was being programmed for half a day before its EEPROM was replaced with

My glamorous wife

another so that, if anything went wrong, only half a day's programming effort could be lost. These EEPROMs were continually being re-programmed until the process was complete. The problem was that the EEPROMs had gold plated legs and every time they were removed from their holder, some of the gold would be scraped off and base metal (steel) would be exposed. Small amounts of corrosion then led to sporadic interruption of the electrical circuits. The reason why the CEGB hadn't experienced the same problem was that, after the entire programming was complete, they'd transferred the configuration on to a brand new EEPROM. Very wise and a lesson learned!

My alarm clock used to be set for 6.30am from Monday to Friday so I could be in Monsanto's office soon after 8.00am and home again by 6.00pm. It was therefore a great joy not to set the alarm at weekends but that didn't mean I got a lie-in. Karen and Michelle would love to come running into our bedroom and climb in to bed with us for cuddles. One morning, Karen announced that she was going to have our bed when she was married, to which Joan asked where we would be sleeping? Karen's very quick answer was "You'll be dead by then!" Charming, we thought. Karen soon became proficient with the operation of our TV and, thereafter, they'd creep down stairs to watch early morning weekend television programmes, which was where we'd find them when Joan and I came down for breakfast. Alternatively, we'd come down to an array of chairs, coffee table and anything else that wasn't nailed down all covered with sheets and blankets. "We're playing tents" would be the answer to our query.

Michelle, me, Willow (pet cat) & Karen

Karen and Michelle were growing up fast and we decided to give them as good an education as we could by privately educating them. Joan's great grandfather, Albert Murphy, started a company in 1887 (eventually called Murphy and Son) which was still operating when Albert died and, in his will, he included a trust fund for his descendents. Joan was a beneficiary of this trust fund and its income was more than enough to pay for our daughters' education. Karen started at Fonthill Lodge in 1980 and was followed by Michelle in 1983. Both girls went on to Notre Dame in Lingfield for their junior and secondary education.

Just when I thought I could contemplate staying with the company for a very long time, Monsanto in the USA decided to re-evaluate what business areas it wanted to operate in and which it would sell off. Unfortunately, the result was that the majority of their European sites would be sold off and they also decided that they didn't need a European Central Engineering Division any longer. The head count was to be reduced from around 300 to just 18 and those remaining would work for what became the Engineering Services Group. Luckily, I was one of the 18 to be retained but the type of work I was to be involved with meant even more travelling. The various sites now became responsible for their own projects and, if they were bigger than they could handle with their own staff, outside design contractors were brought in. My role was to act as a control systems consultant, advising and working alongside site personnel. It was not unusual for me to fly to Belgium on a Monday morning, on to Manchester on Wednesday and back to London on either Thursday or Friday. I didn't find this new role as rewarding as previously, as I didn't think the sites really wanted a consultant from London advising them on what they should be doing. (I'm sure I'd have taken a similar view if I'd been in their position). Whilst travelling from site to site, I often met up with an engineer from St Louis who was responsible for the implementation of business computer systems for all of their European offices. He was based in the European Headquarters in Brussels and on a two-year assignment. Towards the end of his assignment, he spent a great deal of time trying to persuade me to take his position when he returned to the USA. It would have been an upward move within Monsanto but would have involved moving my family to Belgium and the amount of travelling I was involved in would not have reduced. Joan was willing to support the career move but I could tell she would have preferred to stay in England. It was flattering to have been considered for the position, but I eventually declined it.

The next 'bombshell' was Monsanto's decision to relocate to an office out of London to a location "close to a major international airport and the main rail network". Although this could have been near to Gatwick airport, other candidates were Heathrow and Manchester. We were told we would be moving within eighteen months. Joan and I decided we didn't particularly want to relocate and I'd always thought about working for myself but hadn't had enough motivation to do something about it, until then! I'd recently installed an intruder alarm system in our house when it wasn't common place to have them and I realised there weren't many companies offering to install them. I decided to start a company, BurgAlarm Limited, and give myself the eighteen months to see if I could make a living out of it. I installed the first couple of systems myself in my spare time and then started subcontracting the installation side of things to local electricians.

Unfortunately, Monsanto found a new office in Basingstoke within a few months of me starting up BurgAlarm and it was not clear that the business would be big enough to support me, at least in the short term. Basingstoke was far enough from Copthorne to not be commutable (there was no motorway between the locations in 1984) so we started looking for possible houses and schools near the new office. April Cottage was a very nice home and we couldn't find anything comparable to it. It was crunch time: should I relocate to Basingstoke or look for another job? I was put under pressure by Monsanto to decide whether or not to relocate, even though the move would not be for another six months. In the end I decided that the job satisfaction wasn't what it had once been, we had a very nice home with good schools available for Karen and Michelle, and so I decided to find alternative employment.

GLOBAL ENGINEERING

In retrospect, I should have spent longer looking for another job with a company of the stature of Shell, BP or Glaxo but when I got an offer from Global Engineering, a design contractor working in the oil and gas industry, with its office in Sutton, a half hour drive from Copthorne, I accepted it. Although I had started my career with BP, I didn't think I had enough experience of the oil industry to get a job with a major oil company, and so Global Engineering seemed to be a good way of getting it.

I'd only been with Global Engineering for about six months when I visited my first oil platform (Montrose Alpha) in the North Sea to resolve a flow metering problem that had developed. I flew to Montrose Alpha via helicopter from Aberdeen and, once on the helideck, I couldn't believe how cold and windy it was. It was so cold that nobody was allowed to work in the process equipment areas for more than 30 minutes at a time before returning inside for a further 30 minutes to warm up again. Montrose Alpha was one of the early platforms to be installed in the North Sea and its facilities were somewhat limited. The cabins were about 4m x 3m and contained four bunk beds, and a shared toilet and shower with the next door cabin. There was one recreation room for the whole platform where we could watch TV or a film. At the back of the room, there was a pool table, a dart board and a table tennis table but there was only space enough for one of them to be played. Most of the workers smoked cigarettes and thus it was a most unpleasant place to spend an evening. Luckily, I was only there for a couple of days and I certainly wouldn't have enjoyed a permanent job on a platform!

I visited many North Sea and Irish Sea platforms, plus onshore sites at Bacton and Easington during my time with Global Engineering, and also travelled to client offices in Oman and Norway, but my travelling was much reduced when compared with Monsanto. I also didn't have the long commute into central London by train, which was a further bonus.

Karen and Michelle were both thriving at school and took an active part in many extracurricular activities including swimming, gymnastics, ballet, Brownies and Girl Guides. Parent/teacher meetings were always interesting, especially one at Fonthill Lodge where the teacher explained that Michelle was very helpful and always knew who had dropped their pencil at the back of the class, even though she sat in the front row!

One piece of Karen's maths homework from Lingfield I remember related to imagining what a hollow cube would look like if deconstructed. The answer she gave was as follows:

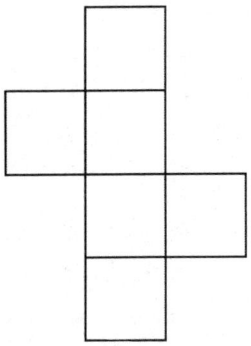

When Karen got her homework back, it had been marked as wrong so she asked me if I could explain why. I knew it was correct and suggested she tell her teacher that I'd confirmed to her that it would fold up into a cube. Back came the teacher's response that she'd been a teacher for X number of years and she knew it was incorrect. I was by now incensed at the teacher's attitude and so cut out the shape, folded it up and asked Karen to give it to her. Karen was terribly embarrassed about the whole thing and wished she'd not said anything to me but, to her credit, did as she was asked. Back came the response that Karen had been marked wrong because her sketch had not been to scale. Teacher wouldn't acknowledge she'd made a mistake but gave this poor excuse. There was a parent/teacher meeting the following week and, although I'd decided not to say anything about the incident (mainly because I didn't want Karen to be victimised), the teacher (now nicknamed as Mrs Cuboid) wasn't looking forward to the meeting and talked nonstop during our 5 minute period with her. She clearly didn't want there to be a pause where I could have asked her about this particular piece of homework.

Mike Clarke, my bridge partner, was married to Helen, and they had two children, Ben and Alicia who was a classmate of Michelle at Lingfield. They had a swimming pool in their garden, which was greatly enjoyed by us all when we visited them in Merstham. We had such fun that we decided to also have one built in our rear garden, and so a major building programme started. The pool supplier recommended a heat pump as the means by which we kept the water warm enough for swimming but I really didn't believe we could get sufficient energy from the air in our English climate. It was probably one of my best ever decisions to also pipe the circulation water through another heat exchanger that was connected to our house central heating system. The heat pump was just about big enough to keep our pool warm enough once it was up to temperature, but it would have taken several weeks to warm it up in the first place every spring season. With the additional help of our central heating system, we could heat the pool water in just a couple of days. Once installed, our pool was used most fine days between May and September and we had many parties and barbeques in our garden.

We had a holiday in British Columbia, Canada in the summer of 1990, and started by visiting Joan's aunt Pam and uncle Sandy in Abbotsford. Freda and John were also there, touring in a camper van whilst our family stayed in hotels. Whilst staying on Vancouver Island, John and I were taken on a fishing trip just off the coast. After about two hours of catching nothing at all, my line went suddenly taught and a fight started. Eventually, I landed a salmon that weighed 12kg and was over a metre in length and, soon after, John landed one that weighed 7kg. I took my fish back to the hotel where we were staying and asked if they would cook it for the eight of us who were having a dinner together that evening. The chef asked what I wanted to do with the rest of the fish as it was too big for us to consume that evening and I offered to donate it to the hotel. In that case, the chef announced, the meal would be free; a very reasonable agreement! As we were checking out the following morning, the chef approached us and handed over a 'cool bag' full of salmon chunks that he'd smoked overnight. We then had smoked salmon for lunch for the next week until we were all thoroughly fed up with it, even though it tasted delicious.

A few days later we tried our hands at white water river rafting! It was really exhilarating and, although we didn't fall in, we were all soaking wet by the time we disembarked the boat. Michelle was very aggrieved because she had to sit in the centre of the boat whilst Karen was allowed to sit astride the inflated side and paddle. Michelle thought she was old enough (at 12 years) to paddle but the organisers didn't agree! Once we were back on dry land, we used Freda and John's motor home to dry off and change into fresh clothes. Unfortunately, Joan inadvertently left our car keys in the boot after she'd collected our dry clothes and we didn't have a spare set. We rang the hire car company to explain our predicament and they told us it would take several hours before they could arrange for someone to attend and drill out the lock. Our hire car was a Ford and, at that time, they prided themselves on their security, so drilling the lock was, apparently, the only option. We were all feeling somewhat subdued at having to wait around when John produced a screwdriver and suggested I might be able to use it to gain entry. I didn't want to appear ungrateful at his suggestion so aimlessly poked the screwdriver at the lock mechanism for a few seconds before 'ping' and the boot lid opened. There was nobody more surprised, or thankful, than me at how easily it had opened. So much for Ford's high security locks! We visited many picturesque towns including Banff, Jasper and the Okanagan valley and did, briefly, think we might emigrate there. However, good sense prevailed and we returned home with good memories of a wonderful holiday.

By 1991, I'd been promoted to Chief Instrument

Freda, Joan, Karen, me, Michelle and John

Engineer and then Chief Instrument and Electrical Engineer responsible for a department of about 20 engineers with technical responsibility across all projects. This position was mainly one of management and I missed the technical aspects involved with project detail design work. All seemed to be going well until I was required, together with a number of others, to visit an onshore oil installation site in Nigeria. One of my contract electrical engineers had returned home from this site a few months earlier with a rather nasty infection that kept him off work for a number of weeks. With that in mind, and as I was the most senior engineer involved with the visit, I approached Engineering Manager, Colin Moss, and asked for the company to insure us against becoming ill and not being fit to work (we were all contract engineers who were only paid for the hours we worked). Global Engineering refused (on the grounds that we were responsible for our own insurance but that was limited to only working in the UK). We all declined to go to Nigeria without either the company continuing to pay us if we became ill or paying for suitable insurance. Unfortunately, I was seen as the 'ring leader' and my contract was immediately terminated (contrary to the contract which stated four weeks' notice). The company subsequently paid for insurance for the rest of the group who went to Nigeria.

I was incensed by this and sought legal advice before taking the matter to court. The judge agreed with me that Global Engineering had no right to terminate my contract in the way it had and awarded me compensation. In his judgement statement, the judge commented on Global Engineering's unacceptable and cavalier approach to the health and safety of their employees. I felt my actions had been completely vindicated!

GENESIS

I had quite a lot of contacts in the oil and gas industry by 1992 and was invited to join Genesis Oil and Gas Consultants by their MD, David Rayburn, who had been a client engineer overseeing a project whilst I'd been with Global Engineering. The only problem with working for Genesis was that I needed to commute to central London again!

In 1992, Genesis was a company that specialised in what was called 'front end engineering design'. That meant they did all of the initial design, usually for client companies like Shell and BP, before the project for detail design was awarded to much larger contractors. However, some of the client companies had been encouraging Genesis to become more involved in the detail design, especially for some of their smaller projects, and thus company expansion was called for. I'd only been there a short while before being promoted to Chief Instrument and Electrical Engineer but still with a role as an engineer and not just a manager. Quite a lot of the work was associated with modifications to existing oil and gas facilities for increased production, more or better automation and changes necessary when the oil/gas fields became depleted. This was more challenging than just designing new facilities (which tended to be fairly routine) and was more satisfying employment.

An early project at Genesis was for British Gas, and we completed the initial design before inviting larger contractors to bid for detail design. Once their bids had been evaluated, three were selected for further scrutiny and I was a member of the team chosen to interview the contractor's representatives. I'm sure you can imagine the look on the face of Colin Moss of Global Engineering when he entered the conference room and saw me there as one of the interrogators! Global Engineering were unsuccessful in their bid and I had a spring in my step when I left the conference room knowing that I'd asked some revealing questions that identified inadequacies in their tender. Et tu, Brute?[2]

Another project I was involved with was for Phillips Petroleum on their Hewitt field whereby the platform was to be modified so that it could be remotely operated from a control room in Great Yarmouth. During the final installation and commissioning, I was required to spend a couple of two-week periods working offshore and, although I'd previously visited many other offshore platforms, this longer stay normally necessitated attending an offshore survival course. Genesis's approach was for me

---

[2] "Et tu, Brute?" is a Latin phrase meaning "and you, Brutus?" purportedly the last words of the Roman dictator Julius Caesar to his friend Marcus Brutus at the moment of his assassination.

to go offshore without attending the course and, if asked if I'd completed one, to say that it was booked for me. I duly started work offshore and, part way through my first period, I worked alongside a specialist engineer who was commissioning a power generator. I saw him at the end of his first day offshore, waiting for the helicopter to take him back to shore. I commented that surely he hadn't finished his work already and he responded that Phillips Petroleum wouldn't let him stay offshore because he hadn't completed the requisite course! I immediately called my office to let them know what had happened and they told me to keep quiet and they'd arrange for me to attend the course as soon as I returned to shore. After I'd been offshore for twelve days, the Offshore Installation Manager saw me and asked me to contact my office and arrange for them to fax my Offshore Survival Course certificate to them for their records. I duly called our project manager who was tasked with admitting to Phillips Petroleum that I hadn't been on the course. I was made to 'shuttle' (travel back to shore for the night and back again in the morning) for my final night even though the most dangerous part of working offshore was the helicopter journey. Phillips Petroleum was making a point!

I attended a three-day course in Yarmouth the following week to learn what to do if a helicopter ditches in the sea, how to fight an offshore fire and how to evacuate a platform by lifeboat. The helicopter ditching was the most alarming, especially as I wasn't a confident swimmer. We were made to sit in a mock-up of a helicopter that was lowered into a swimming pool and we had to get out and into a lifeboat. Next time the helicopter sank, we had to get out when under water. Finally, the helicopter was lowered into the water; it then sank and turned over. We had to wait with seat belts on until we were upside down before releasing the belts, leaving via the windows or doors, and swimming to the surface. All very nerve-racking! However, I passed the course and was allowed to stay on the platform for my next two-week stint.

Another interesting gas processing design project was for Lasmo in Columbia. One of our process engineers was required to visit the site for commissioning, but a few months before he was due to go, a British man was kidnapped in Columbia and this became high profile in the British newspapers. It turned out that lots of people were kidnapped every year but it usually didn't get reported in England. The wife of the process engineer wasn't happy about him going and he was teased about it in our office. However, I was eventually asked to accompany him on the commissioning visit and we asked the company to consider kidnap insurance. The quoted insurance premium was £15,000 each for a two-week trip! Genesis and Lasmo said the project couldn't afford to pay this but, if we went and were kidnapped, they'd continue to pay us for 6 months. Having previously refused to go to Nigeria, and lost my job because of it, I was not inclined to refuse again, so accepted the written confirmation about continual payments if the worst happened. I gave the letter to my neighbour and told him what to do with it if I didn't return! We were also promised that we'd travel to the site by different routes every day, travel at different times and that our vehicles would have direct radio communication with their head office, all to mitigate the chances of kidnap. We travelled to Bogotá in December of 1995 and then on to Sincelejo, the nearest town to the site in Guepaje. The hotel we stayed in was modern and even had a swimming pool. However, the water was green and so not suitable for swimming. The room showers were also interesting; there were hot and cold taps but, in the mornings, cold water came out of both of them. In the afternoons, only the cold tap would supply water! However, the weather was very hot whilst we were there and using cold water was not too much of a hardship. The travel promises didn't materialise as there was only one road between Sincelejo and Guepaje, a distance of about 40km. We left every morning at dawn and returned just as the sun was setting. It was considered too dangerous to travel when it was dark and it was important that we were on site for the maximum time available. However, the Toyota Land Cruisers did have radios for the first couple of days and then they disappeared until we complained about their omission.

Guepaje was a village with a mixture of old buildings and 'wattle & daub' type constructions with mud walls and palm leaf roof thatching. The people who lived in these huts always seemed cheerful even though they were clearly very poor.

In addition to the gas that our site was producing, there was a small quantity of hydrocarbon condensate (a clear oil type substance) produced that was not financially viable as an export product so the villagers were allowed to queue up on a Saturday morning to fill plastic or glass containers with it for burning in their oil lamps. I was somewhat surprised to see these same people leaving their huts on a Sunday morning dressed in their finery and on their way to church. Even though they were poor, they had huge pride in their appearance and dressed up for special occasions.

We returned to England on 23rd December having seen no sign of problems. We later discovered that most of the kidnappings were associated with the drug cartels that operated in the south of Columbia, whereas we were working in the north. With Christmas over and contemplating returning to work in January, the telephone rang on 30th December and Dave Rayburn asked if I'd had a good Christmas. He'd never done that before and I immediately 'smelt a rat'. Apparently, the site in Columbia had developed some operating problems and Dave wanted to know if I'd mind going back to help sort them out. "When?" I asked. Dave suggested "Tomorrow?". I packed my case and was at Heathrow the following morning. The Lasmo staff was appreciative of our return and we quickly sorted out the issues.

OUR EMPTY NEST

I always knew I didn't want to be working in a full time position up to my 65th birthday but was also sure that, upon retirement I wouldn't want to sit around with my feet up. Thus a chance conversation with a local Magistrate sparked a desire to find out more about a possible 'career' other than being an engineer. This led to my visiting the Magistrate's court in Crawley, West Sussex and eventually requesting an application form from them. One of the questions on the form was, 'Does my employer support my application?' I hadn't, at that time, asked, and upon finding out, the answer was a definite 'No,' as "it could be detrimental to the projects I was involved with". I answered the application form question with 'discuss at interview'.

The meeting with the court's Advisory Sub-Committee went well and among other things, I was quizzed about my attitudes towards ethnic minorities and young people and the way they went about life. My interrogators also put imaginary scenarios to me about what they said was a typical case, invited my comments and, then, probed me with further questions to see my response. In answer to the 'employer support' question, I explained that my application was not supported but that

I was prepared to terminate my staff employment if appointed and revert to being a contract employee.

At the end of the interview, I was told that it could be months before any decision was made and I should not be disheartened if I was rejected as even if I met the criteria needed, the Bench had to be balanced – age, gender, ethnicity, social and occupational background etc – and a surfeit of candidates with similar backgrounds might mean that not all would be appointed, so I left the meeting with no expectation of being taken on. So it was with some surprise that soon after my interview with the Advisory Sub-Committee, I received a letter in March 1999 informing me that it was intended to put my name forward to the Lord Chancellor for possible appointment as a JP for Crawley, subject to my confirmation that I had a "clean" record of no convictions or court orders against me. Following that confirmation, I received a letter notifying me that my appointment as a magistrate had been confirmed, almost fifteen months after my initial contact with the court. I was "in" and was about to start out in the most interesting task that any volunteer could ever have. I told AMEC about my appointment, which meant I'd spend about 13 days per year not being available at work, and they confirmed that I could remain with them if I reverted to being a contract employee, and as long as they knew well in advance when I'd be unavailable. No problem! I registered another limited company, Offshore Controls and continued working.

Towards the end of a project for BP, Wytch Farm, my involvement didn't justify a full-time commitment and I was bored trying to fill a day without enough to do. For obvious reasons, I couldn't sit around reading a newspaper or 'surfing the internet', so I requested permission to reduce my working to two days/week. This was rejected and so I brought forward my retirement! The first few months were great and I played a lot of golf, caught up with my DIY backlog of home projects and volunteered for additional magistrate's court sittings. However, by the end of October 2001, I wasn't sure how I would fill my time during the winter months (I was never a fan of playing golf when it was wet and/or cold!). A short time later I received a call from AMEC asking if I was available to help out with a bid they were involved with. I agreed but only on the basis of working two days/week. My predicament about how to fill the winter months was solved as, once the bid had been completed, they asked me to stay on working part-time.

MOVING TO THE COTSWOLDS

Around the year 2000 we and our four immediate neighbours received letters from a building company asking if we'd all consider selling our homes so that the site could be redeveloped. We'd had a similar request in about 1989 but, after initially reaching a generous agreement, the builder reduced his offer because he couldn't get permission to build as many houses as he had hoped. We had been going to accept solely because the offer had been so good and, now that the incentive had gone, we declined the reduced offer. This time we were much more inclined to accept as I was already working only part-time, Karin (her preferred spelling!) and Michelle were no longer living with us, and we felt we should consider moving closer to my 'in-laws', John and Freda White, who were then in their late 70s and may have needed help as they aged. The whole process took more than two years to complete by the time we'd all reached agreement on selling prices, and the builder had obtained the necessary planning permission.

In the summer of 2003, Joan and I started looking in earnest for a house within about an hour's drive of Harbury (where John and Freda were living) and investigated Oxfordshire, Northamptonshire,

Herefordshire, Gloucestershire and Warwickshire. We wanted a house with 'character', good elevated views of the countryside and close to a major town. We rather liked Malvern and there were lots of houses with views, as the town sits high up in the Malvern Hills, but it was too soon to look for a particular house as the building company were still awaiting planning permission for their development. It was around this time, when Joan and I were attending a morning service at our local church in Copthorne, that the sermon caught my attention. The preacher held up a large glass jug which he filled to the top with large pebbles. He then asked if the congregation thought the jug was full. We all did but the preacher then produced a bag of small stones which he proceeded to pour into the jug and they filled the spaces between the pebbles. Again he asked if we thought the jug was full and, when we agreed it was, he produced a bag of sand and poured this into the jug. Was the jug full? We were more cautious this time and declared that it probably wasn't and the preacher then poured water into the jug right up to the brim and then, when confirming the jug to be full, asked what we thought the moral of the story was. There were several suggestions along the lines of no matter how full our lives were, we could always fit in a bit more. No, the moral of the story was that unless you put the biggest items in first, there will never be space for them later.

This made me think about our move and the fact that the church we would attend after moving was very important to us. A couple of weeks later, we spent the weekend in Malvern, attending Malvern Priory on the Sunday morning. We enjoyed the service, felt welcomed and were invited to stay for coffee during which we chatted to several people including someone from New Zealand who was visiting his brother in Malvern. We mentioned that we were investigating local churches and looking for one that was more evangelical than St John's in Copthorne, and he recommended that we try St Matthew's church in Cheltenham. We travelled back to Copthorne via Cheltenham (which we'd never visited before) and liked the look of the town, with its Georgian architecture. We later spent another weekend away, attended St Matthew's and immediately felt completely at home. Joan and I believed we'd been guided to Cheltenham through the initial sermon, and the people we'd subsequently met.

My journey to faith was along a road of many turnings. I went to Sunday school as a child up to about the age of 14, when I rejected religion as a prop for those who needed to have a reason for being here on earth, and the hope of an afterlife. I kept that view until Joan and I were invited by Jackie and Pete Mankelow to go along to their church, St Andrew's, in Crawley when Karen and Michelle were quite young and had started school. The services at St Andrew's were less formal than traditional churches and, although I still wasn't a believer, I enjoyed being there. Over very many years I'd hoped and prayed for a 'road to Damascus' conversion but it never happened, and so I just kept on wondering about who Jesus was, and why he lived 2000 years ago. I've no proof that Jesus was the son of God but there is sufficient evidence of his existence, his life and crucifixion, but did he actually rise from the dead? I eventually believed he did, mainly due to the fact that so many of his early followers, who were actually around when Jesus was resurrected, were killed for their beliefs rather than renounce them. Would they really have done that if it were not true? I believe they wouldn't.

Once we'd decided to move to the vicinity of Cheltenham, we focused on looking for a home within 10 miles of the town and in March 2004, we moved to

Meadow Cottage

Meadow Cottage in the village of Syde, near Cheltenham, Gloucestershire.

The house was originally an old farm worker's cottage that had already been renovated and extended and was located in a wonderful rural location with westerly views over a valley. We knew the accommodation wasn't what we ideally wanted and so employed a local architect and builder to remodel it into what we required.

I also transferred to the Gloucestershire Bench as a Magistrate and sat in Cheltenham, Gloucester and Stroud. I became a Chairman in both the adult and youth courts. I can say that I enjoyed my time on the Bench but the word 'enjoy' can be wrongly interpreted as there was no pleasure in many of the duties that befall a magistrate – better that I should describe the experience as interesting, rewarding, frustrating, enlightening, challenging, provoking and, undoubtedly, demanding. The work of a magistrate brought me into contact with the bad and the sad, the feckless and the devious, the person who makes one mistake and, thankfully, never offends again, and the persistent malefactor who regards court appearances as a way of life.

Before we relocated, I had moved back to Genesis but was still just working two days/week. I didn't intend travelling to London every day, but offered to continue working for them as a design consultant, based at home. This arrangement worked well for both of us, as I didn't have the hassle of commuting, and they benefited from my being more productive when working away from the office. I also worked for a number of other clients as a consultant, working mainly from home, with the occasional office meeting or trip to a site. I still enjoyed the intellectual challenges that came my way and was eventually persuaded to do some part-time 'Expert Witness' assignments. These were associated with examining oil and gas projects that had overrun financially and where project completion was delayed. The client companies were trying to claim damages from their design contractors whom, they considered, had caused the overruns and delays. In both projects I worked on, I was generously paid by the client companies and enjoyed the work for the first 6 months whilst evaluating the merits of the claims. Thereafter, and for about another 18 months on each project, the work became extremely tedious whilst writing, rewriting and generally massaging the reports I'd prepared in the initial six months. In 2012, after two projects, during which I'd earned a lot of money, I considered it was time to finally 'hang up my working boots'.

I joined Cheltenham Bridge Club after our relocation and was quickly elected onto the committee to organise the internal competitions and publicity. In 2011, I was elected as club Chairman and led the club through a major refurbishment programme of the building that the club owned, as well as a teaching programme to bring in some younger blood. After my two year tenure as Chairman, the club membership stood at over 500 and was thriving. I won many competitions at the club and my name is on several of the Honours Boards. I also represented Gloucestershire on a frequent basis in inter-county competitions and reached an English Bridge Union rank of 'Life Master'. It's true to say that the game of bridge became an important leisure time activity for me, especially once I'd retired, and I thoroughly enjoyed the mental challenges necessary to play the game well.

Karin was working for Matrix in London from 2002 and was living in Barking. She'd had a number of lodgers staying at her house whenever she wanted to save up for special purchases or holidays and the latest one was John Campbell who was on secondment from Wellington, New Zealand. John turned out to be a very special lodger!

Joan and I travelled to New Zealand in January 2010 for Karin and John's marriage where we met up with our Michelle and her partner, Andrew Cyril. We then travelled by the inter-island ferry from Wellington to Picton, and then on to an idyllic small luxury lodge hotel at the Bay of Many Coves, Queen Charlotte Sound, Marlborough. Karin and John hosted a BBQ for guests to enable the families to get to know each other before they were married on 23rd January 2010. Following the wedding, Karin, John, Michelle, Andrew, Joan and I travelled along the North coast of the South Island to Nelson for a few days, where we enjoyed the tourist attractions, including a great kayaking expedition in the Abel Tasman National Park. Karin is currently the Director of Strategic Performance at the Ministry of Justice, New Zealand.

Michelle was working as a paediatric clinical nurse specialist in London for University College London Hospital when she met Andrew Cyril who also worked there. They decided to get married in August 2012 giving just six weeks for the wedding preparations. They were married on 22nd September 2012 at St Mary's church in Syde, Gloucestershire. St Mary's is a very small church and we filled it to capacity with friends and families. The weather couldn't have been kinder, warm and sunny, which was just as well as the reception was held in a large marquee in our garden. Regrettably, Andrew died from a brain tumour on 17th April 2014 and his remains are buried in the churchyard where they were married.

I'd been having PSA (Prostate Specific Antigen) blood tests from about the age of 50 to check for possible early signs of prostate cancer and in 2013 I was told by my GP that my PSA level had increased. My doctor referred me to Gloucester Royal Hospital for a prostate biopsy but, before my appointment, I'd done some internet research and found that University College London Hospital was involved in a clinical research programme for better detection of prostate cancer. I asked to be considered for the trial and, fortunately, was accepted. I say 'fortunately', because in addition to the new tests, they also conducted the standard test which missed the fact that I had a cancer. I was treated in 2014 using a 'Nanoknife' (also known as electroporation), which was also being trialled at UCLH. This involved using a number of needle type electrodes to surround the tumour and then pass low voltage current between them to destroy the cancer cells. Later MRI scans and a biopsy confirmed that the vast majority of the cancer had been destroyed.

I'm now 68 years old and have been rambling on for longer than I intended. I've been fully retired for three years and I still don't have much free time although I no longer feel guilty if I sit down and read a book in the afternoons. I'm still working as a lay magistrate, I'm the treasurer of our local church and on the executive committee of Cheltenham Bridge Club. My leisure pursuits include membership of various U3A (University of the third age) groups, badminton, bowls, bridge, gardening, visiting National Trust properties and walking in our beautiful countryside but not forgetting frequent holidays and weekends away from home. I still attempt DIY jobs around our home but am also prepared to leave some things to those more proficient than me. My family gave me the nickname of "Mr Fixit" many years ago as I've always been willing to attempt repairs and generally get things working properly.

I can look back on my life and reflect that I'm privileged to have lived in a period without European wars and there have been precious few generations who have been so lucky. I've travelled extensively, for leisure and work, which has been rewarding and stimulating as well as relaxing. Perhaps some of my life-directing decisions were questionable at the time, such as resigning from BP, not relocating to Basingstoke with Monsanto, and refusing to go to Nigeria; if I'd gone to Grangemouth with BP, then it's unlikely that I'd have married Joan (my wonderful wife and soulmate); if I'd stayed with Monsanto, then I'd

Elaine (left), me, Brenda and Mum

have seen a lot less of Karen and Michelle whilst they were growing up, and if I'd travelled to Nigeria without insurance, I'd have let Global Engineering get away with neglecting their responsibilities towards all employees.

My career as an instrumentation and control systems engineer was ideally suited to me. I got immense satisfaction from designing controls to enable processes to operate automatically and subsequently commissioning them once the facilities had been built. I've consider myself to have been a dedicated employee who always contributed to the best of my ability but I also saw work as a means of earning enough to enjoy my leisure time. I was conscious of my need to achieve a good work/life balance and tried to avoid spending too many hours at work. This wasn't always possible during my time with Monsanto when travelling to various locations was necessary. Not working too much overtime enabled me to be a true 'family man' and I've enjoyed having lots of time with Joan, bringing up Karin and Michelle, and giving them the best start in their lives.

I can truly say that I've got absolutely no regrets about the decisions I've made throughout my life and I look back on my achievements with pride and satisfaction.